Architectural Drafting

in

MiniCad 6

Jon Stoppi

Architectural Drafting in MiniCad 6

Qualum Publishing
665 Finchley Road
London NW2 2HN, UK

British Library Cataloguing-in-Publication Data.
A catalogue record for this book is available from the British Library

Cover design: Yvette Yanne
Certain images in cover illustration courtesy of Geoffrey Bowman Architect, Sigma Design
All other illustrations by the author.

Printed and bound in Great Britain by the Bath Press Group plc
ISBN 1 899 168 133
Price: £29.95 (US$45.00) nett

Andy Branfan 16.12.96.

Introduction

In MiniCad 6, Graphsoft has once again done a good job of producing a program that looks deceptively like its predecessor but tucks away a wide range of innovations and improvements in various nooks and crannies of the same basic interface. This book was started with the same intention towards its previous edition, but has ended up bearing roughly the same relationship to it as *Godfather II* did to its original: the name's very similar, but it's a new storyline, new faces—and this time there's no horse heads in it.

More specifically, the book has been completely rewritten and revised, employing a more elaborate and realistic sample project (without wearying the reader with more than is necessary to cover the various features) and above all providing a comprehensive discussion of the program's 3D features in an architectural context.

Nevertheless, the basic format, is the same as before. The reader is not presumed to know anything of MiniCad beyond the basics of Macintosh operation*, and is led to decipher its operation through a familiar context and associative sequence of solutions to specific issues along the way. Readers of the previous edition may recognize the basic scenario in the beginning, but will soon appreciate the differences arising from the integration of the program's 3D features as well as those that are new to this version of the software.

A further change is that on this occasion I have gone to some effort to produce a 'universal' edition, i.e. one that is equally applicable to both the North American and metric markets. To this end, I have used a model familiar to (if not typical of) domestic architecture throughout the English-speaking world, and minimized reliance on country-specific or even climate-specific features. Furthermore:

- Issues specific to the (North) Americ*a*n readership are marked @; those for the benefit of readers in the Rest of the World with ®.

---

* To all you users of the new Windows™ version: welcome aboard, and apologies. A separate edition is planned for your benefit later this year. However, if you can reconcile the different keyboard names and Mac screenshots, this book can serve as well, as the application is essentially the same.

• All dimensions are expressed in both English (Imperial) and metric units, with likely equivalents rather than precise conversions being the rule (thus, 8000mm is 26'-3", not 26'2.961").

• Most of the drawing on my part was carried out in metric, but detailed wall construction, stud dimensions etc. are based on the two-by-four family.

• Since the program and most software terms are American, and Americans tend to be more distracted by British spelling than the other way around, American spelling rules (with the exception of the metric [®] references). To make it up to my fellow 'aliens', however, where figures must be shown in screenshots of dialogs these will often be in metric, and the style of writing is British (or so I'm told).

A Guide to Graphic Cues & Typefaces

Like its predecessor, this edition emulates the page size, and basic cover binding design of the Graphsoft manuals so as to complement them naturally on the shelf. Inside, however, everything is different, and the reader may find it useful to note in advance that, while the main text is in Palatino 10

> *Incidental comments that are not part of the tutorial procedure but worth knowing are in Palatino 9 Italics like this…*

 Note: *…While particularly important points to note are flagged like this.*

> Where such explanations warrant more than a line or two of text, they will often be boxed off in a section of their own, surrounded by a dashed line, and typeset one point smaller than the main text.

For the sake of clarity (and to avoid excessive use of quotation marks which often confuse more than they help) references to words and figures on screen will usually mimic their appearance there in terms of typeface. Thus, menu items and dialog text that appear in **Chicago** will be printed as such here, too. Screen hints and palette items will similarly be set in **Geneva Bold** or Plain as appropriate. In addition, keyboard references—i.e., buttons or words that you type that don't result in text on screen—will generally be in `Courier`.

For the benefit of veteran users, features new to MiniCad 6 or which have undergone significant changeses will generally be flagged as such.

Finally—although it should be fairly obvious—where this is helpful a ⚫ ▶ gray arrow will demonstrate the action discussed in the screenshots, indicating both direction and initial and intermediate click-spots (spots without lines indicate clicks without dragging)

Acknowledgements

I am deeply indebted to my friend Rafi Shafir for advice on some of the differences between architectural practice on either side of the Pond, and to my family for its patient support as always. In addition, my thanks to:

• All those at the coalface of day-to-day professional MiniCad practice who responded to my call earlier in the year for insights on the extent to which various features are actually used and other comments. I refer in particular to Vegard Brenna, Petri Sakkinen, Jonathan Pickup, and Dave Weber (whose updated *DXF Made Easy* I mention here but have not included in this edition, given its availability on the now-ubiquitous World Wide Web)

• All you readers and clients who were so generous in praise of the previous edition and fervently spurred me on to produce this one and bore patiently with me as I struggled to complete it in the midst of other commitments

• Last but far from least, to the indefatigable TeAntae Turner—the rock of Graphsoft Technical Support—for prompt and pertinent responses to my endless questions on various undocumented, not-yet-implemented, and... er, other features.

'Nuff said. Hope you find the book useful.

—JS
August 1996

§

Contents

cont./...

(.../Contents)

(.../Contents)

§

Installation

Insert Disk #1 of the set that comes with your package.

The following window should appear
(if not, double-click the disk's icon to open it):

Double-click the Installer to launch it. You will get an intermediate window welcoming you to the MiniCad 6 Installer, with a single button option—**Continue**—which you should click to get to the important screen:

I recommend *not* going for the default option (**Install** – here, as always, the option encircled with the bold line), as this installs what is called the 'fat binary' version of the software – no less than 31Mb, as you can see – on your hard disk. This is in fact two versions of the application: one for PowerMacs, and one for 68k-types. It is useful if you are installing on an external hard disk which may be used alternately with machines of either type, but in all other cases you will probably prefer installing the version for your particular machine. So click **Custom**

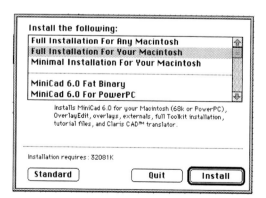

…and choose the second option from the top: **Full Installation for Your Macintosh**.

This adapts the installation to your particular machine. Follow the instructions on screen: you will be asked for each disk in turn of the set that arrived in your package, ending up with the first disk to complete the installation.

First Acquaintance

The very first time you launch the program, your opening screen should look something like this:

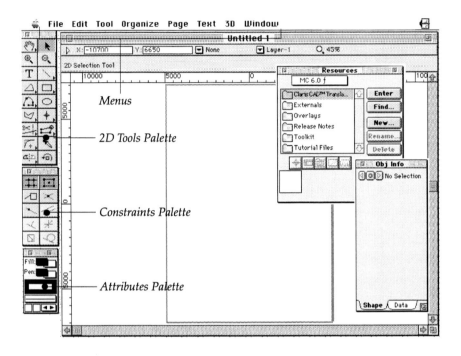

This is the **Standard Overlay** of the program, and you are not limited to this particular arrangement of tools and menus if you don't like it.

> *Veterans of previous versions will notice that the buttons and Data Display Bar have a new, 3D-like appearance that is all the rage these days, but are otherwise much the same.*

> *Incidentally, ignore the* **Resources** *and* **Obj**(ect) **Info** *palettes for now: we will come to them later, and in fact you should close them by clicking on their close boxes in the upper lefthand corner.*

Other Overlays are available and you can even customize your own, using the **Overlay Edit** utility that comes with the package. You may not feel ready to do any customizing just yet, which is why, for those of us wishing to access the program's architectural features, a special

AEC Overlay is available to us from the list of preset overlays.

This appears as a sub-menu of the **File>Overlays** menu item:

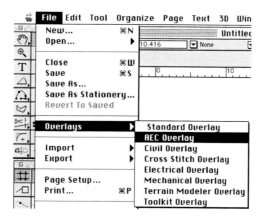

After choosing **AEC Overlay** from this submenu,
the tool palettes will disappear for a few seconds,

and when the new Overlay is in place you will notice that the main (2D) palette has a few added tools, and a new palette for 3D objects and operations has appeared: if your screen is large enough, it will come under the 2D palette; if not, it will 'hide' behind it.

§

Setting up for the First Time

Page	Text	3D	AEC
Normal Scale			⌘3
Fit to Window			⌘4
Fit To Objects			⌘6
Save View...			
Set Grid...			⌘8
Set Origin...			⌘9
Guides			▶
Scale...			
Units...			
Drawing Size...			

The First Five Operations

There are certain other operations that we need to carry out before starting work. MiniCad rightly makes no assumptions as to the type of design intended, so the page size is a single US Letter and the default scale is 1:1. To change this, first choose **Scale...** under the **Page** menu

(The ellipsis [...] always signifies that a dialog box will follow)

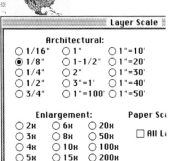

In the dialog that follows:

Layer Scale

Architectural:
- ○ 1/16" ○ 1" ○ 1"=10'
- ● 1/8" ○ 1-1/2" ○ 1"=20'
- ○ 1/4" ○ 2" ○ 1"=30'
- ○ 1/2" ○ 3"=1' ○ 1"=40'
- ○ 3/4" ○ 1"=100' ○ 1"=50'

Enlargement: Paper Sc:
- ○ 2ᴋ ○ 6ᴋ ○ 20ᴋ
- ○ 3ᴋ ○ 8ᴋ ○ 50ᴋ   ☐ All Lᵢ
- ○ 4ᴋ ○ 10ᴋ ○ 100ᴋ
- ○ 5ᴋ ○ 15ᴋ ○ 200ᴋ

⬅ [@] look to the left if you work in the North American system of feet & inches and (for now) choose 1/8"

➡ [®] in the Rest of the World look to the right if for metric/decimal settings and choose the **1:100** 'radio button'

Engineering:
- ○ 1:1 ○ 1:25
- ○ 1:2 ○ 1:50
- ○ 1:4 ● 1:100
- ○ 1:5 ○ 1:200
- ○ 1:10 ○ 1:500

cale: 1: `100.000000`

Layers ☒ Scale Text

[Cancel] [**OK**]

- *The [@] and [®] will henceforth distinguish operations for North America and the Rest of World, respectively*

- *Being an American program, the heading **Architectural:** assumes only Engineers use decimal scales. Note that even when you choose a feet & inches type scale, the **Paper Scale 1:** field on the right always displays the decimal equivalent*

- *In your actual work, if you don't see the scale that you need from the list of radio buttons (e.g. 1:20), just type it (i.e., **20**) in the **Paper Scale 1:** field*

Disregard the **Enlargement** option for now: this refers to the single-step effect of using the zoom tool, and there is no call for it right now.

§

The next step is to choose your **Units of measurement**. By default, these are Feet & Inches: if you need otherwise, choose **Units...** from the same **Page** menu we used earlier.

You get the following dialog:

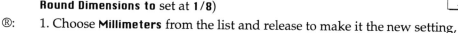

Click and hold on the Unit Name pop-up field to see the list of preset options. Note that the current setting is always marked with a checkmark. For now,

@:   keep all options at their default settings
      (**Feet & Inches**, **Display as Fractions** checked,
      Round Dimensions to set at 1/8)

®:   1. Choose **Millimeters** from the list and release to make it the new setting,
      2. Turn off (deselect) the **Display as Fractions** option,
      3. Set **Round Dimensions to: 0**

In all cases, leave the **Decimal Format** and **Angular Accuracy** at their default settings

> • **Rounding** *only affects the <u>display</u> of dimensions—not the accuracy with which they are calculated (that is set under* **Preferences...** *—see below)*

> • **Old-Style Feet & Inches** *is solely for the benefit of users of early versions of MiniCad, who wish to import and update old files: don't use otherwise.*

The next operation under the **Page** menu is **Drawing Size**. This has been extended and made more relevant to the real world of drafting, by giving direct access to the standard ISO A, ANSI and US Arch sizes through the **Size** pop-up menu. You can also type in the size in mm or inches, and – a great improvement – opt to show or hide the page breaks in the drawing where your printer or plotter has to divide up the drawing into 'tiles' of smaller sizes. Choose

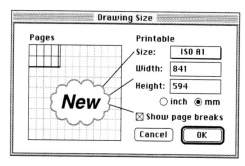

@: **US Arch D**
®: **ISO A1**

The size of the 'tiles' you see depicted in this dialog are determined by the type of output device you have selected in the Chooser. If it is a plotter, the plotter driver will have its own characteristic Page Setup dialog, enabling you to get the whole of the chosen size drawing on one tile. If it is a laser or dot-matrix printer the maximum sizes will be more limited. In any

event, you make your choice of size from the pop-up menu and of orientation from the appropriate icons.

Our final step at this point is to set our **Preferences...**. This is under the **File** menu

New

A logical and welcome change from previous versions

This is where many important aspects of the drawing operation are set, and there are many things to note here, most of which are self-explanatory and can be left at their default setting. But for our purposes, make sure:

• **Click-drag drawing** option is *off*, and
• **Offset duplications**, **Screen Hints**, **Use floating datum** and **Eight Selection Handles** are *on*.

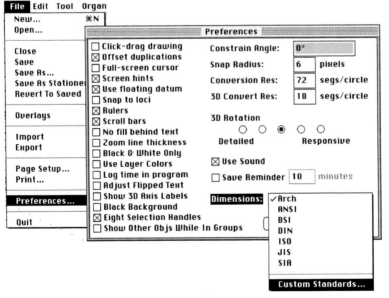

Note, too, the Dimensions pop-up field for choosing your preferred convention of dimensioning, including those set out by the national standards of the US (**ANSI**), the UK (**BSI**), Germany (**DIN**), etc. Choose the one your prefer, leave at the default of **Arch**, or make your own Custom Standard by choosing that option at the bottom...

In the dialog that follows,

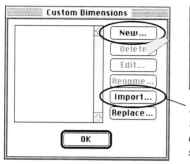

Note that you can import a standard that you have already made in another file: this is a theme that we shall return to in a moment

click on the **New...** button

and give it a name, **OK** the dialog, and…

select the newly-created standard and click on **Edit...**

Note the many options available to suit your most precise requirements as regards type and distance of dimension and witness lines, text, orientation of text, markers, etc.

Click on the **Select...** button to change the default markers from the defaults

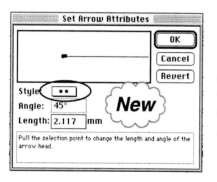

shown to one of the others available in the **Style** pop-up in the **Set Arrow Attributes** dialog. This gives complete control over the size, angle, and appearance of slashes, circles or other markers you may choose.

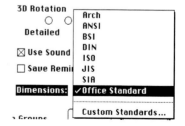

Finally — a common omission — don't forget to choose your customized Standard from the list back in the **Preferences...** dialog!

The final operation now is to save our file. Normally, this would be done by simply pressing Command-S or choosing **Save** from the **File** menu, but on this first occasion we want to **Save as Stationery...**

As in other programs, stationery files save your particular combination of settings so that you don't have to go through the same hassle of changing the default settings to your liking each time you create a new drawing file. You will typically create a series of stationery files — one for each kind of size drawing, for example — and then use *them* to launch the application instead of the application icon itself. So far the sort of settings we've made are fairly basic, but as you learn the various features of the program you will be able to (and should) add these, too, to the stationery files, so that all your favorite preferences, commands, symbols etc. are all at hand.

Whenever you ask to **Save as Stationery...** MiniCad suggests the name **MiniCad Default**: accept it, because a file of this name in the application's folder ensures that these settings become the default settings for the program whenever it is launched on its own (there may be an existing file by that name already, in which case, confirm that you wish to Replace it with the new one). But I recommend saving a second copy in a separate folder — perhaps one designated for your future collection of

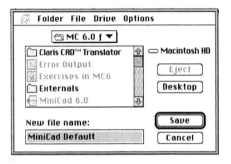

MiniCad stationery — under a more meaningful name that gives information about it, e.g. **@: DL1/8"** (= Arch D size, Landscape orientation, 1/8" scale) or **®: A1L/100.mm** (=A1 size, Landscape orientation, 1/100 scale, mm units).

Stationery file icons look different to their standard counterparts so you can tell them apart straight away, but I sometimes also add a ™ to the name as a kind of private code for 'template'— this helps me identify stationery files at a glance, and to find them easily through the Find File search routines.

This has become more important since, unlike in previous versions for some reason, the icons of stationery files now no longer look different in the MiniFinder dialogs such as the one above.

New

Having made the Stationery file, close it or quit the program, then double-click on it to create an **Untitled** file, and **Save** that (Command-S) to practise with. Call it, say, **MiniCad – The Drawing**.

Drawing Tools

The process of drawing in MiniCad — unlike some PC CAD conventions, but similar to all native Mac draw-type applications — involves first choosing (clicking on) a tool from the palette, then click-and-dragging or click-move-and-clicking in the drawing area to mark the start and endpoints of the object in question.

On both the 2D and 3D palettes the tools are of three kinds:
- navigation or manipulation of the drawing/model
- object creation or placement
- object manipulation and editing

These are arranged into more or less logical groupings as shown below (more so in the 2D palette than in the 3D). The little arrowhead at the bottom right of some of the tools means that other, related tools or functions are available on a 'pop-out' palette if you click and hold on it.

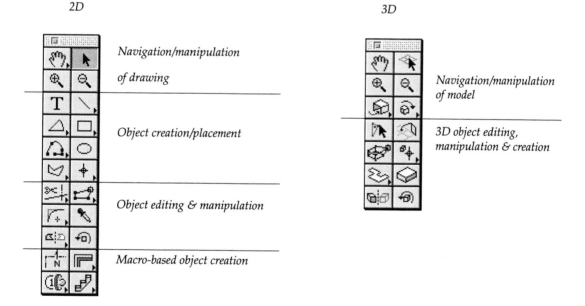

Experiment using the 2D tools (we'll come to the 3D ones later) to get a feel for the general technique. An in-depth description on each tool is available in program's Reference manual, but we will also examine them in turn as we go along.

The **Constraints Palette**—as befits its name—allows you to constrain your drawing operations in a number of ways:

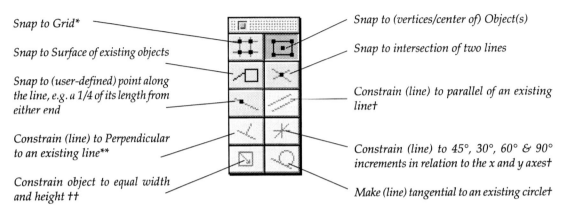

Snap to Grid*

Snap to Surface of existing objects

Snap to (user-defined) point along the line, e.g. a 1/4 of its length from either end

Constrain (line) to Perpendicular to an existing line**

Constrain object to equal width and height ††

Snap to (vertices/center of) Object(s)

Snap to intersection of two lines

Constrain (line) to parallel of an existing line†

Constrain (line) to 45°, 30°, 60° & 90° increments in relation to the x and y axes†

Make (line) tangential to an existing circle†

These constraints are often used in combinations, e.g. Snap to Grid and Snap to Object (the default setting). They are particularly handy when you know how to call them up 'on the fly', using one-key presses (no Command key necessary) on the lefthand side of the keyboard—leaving the right hand free to draw with the mouse. The most frequent ones are Snap to Object (press A), Snap to Grid (Q), and Snap to Surface (S).

All single-key selectors are detailed in the first panel of the MiniCad Help... file (Balloon menu). They are good not just for constraints, but for tools such as Text, the Selection Tool, etc. I find the ones indicated particularly worth remembering.

By the way, MiniCad does not implement Balloon Help

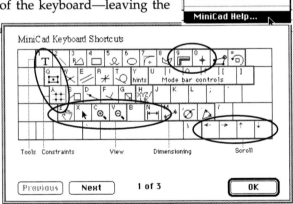

---

* To determine what that Grid is, double-click on that tool, press Command-8 or choose **Set Grid...** from the **Page** menu

** Double-click on the tool to get to its settings dialog

† Enabled only when drawing single lines, single-line polygons and radial arcs

†† Enabled only when drawing rectangles or ellipses

Getting a Handle on Selected Objects

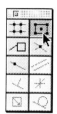

The Snap to Object constraint is worth going into in some detail as it offers an important insight into how MiniCad (and other draw-type applications) describes objects. MiniCad makes the distinction between two basic types of graphic object:

- Lines
- Surface objects, or objects that describe an area

Line objects are simple: they are described mathematically by just two points— their start and end, respectively.

Circles made by one of the circle options now have only two handles: but the snap points are still at the bounding box handles

Surface objects are basically everything else: rectangles, ellipses & circles, arcs of all kinds, bezier curves and freehand lines, polygons and polylines. The **bounding box** of an object is the imaginary rectangle that just encompasses the entire outline of the object.

In the case of ordinary rectangles this coincides with the outline of the object itself, but not of course with other objects.

Freehand objects, arcs & polylines are also surface objects

If we were to draw these objects with a visible fill and then select them, we could clearly see how this works for the various types.

In most cases, Snap to Object means that you will snap to the points or vertices of the object. However, with objects with no vertices such as rounded rectangles, circles, ellipses, non-radial arcs, and bezier curves, what you are really snapping to is to the eight little black square or 'han-

cont./...

.../ cont.

dles' at the corners and the middle of each side of their bounding boxes. A ninth snap-point—which is invisible—is the center of the object. The exceptions to this rule are radial arcs (snap points at each end of the arc, plus one at the arc's center; the one in the center of the arc is not a snap point).

The nine points of the bounding box of an object and its centre are also used to establish the position of an object in various dialogs. Draw any object with a surface area — e.g., a rectangle — and then open up its **Obj**(ect)**Info** palette (**Window** menu; or press Command-I).

Notice the miniature representation of the object's nine handles, one of which is selected.

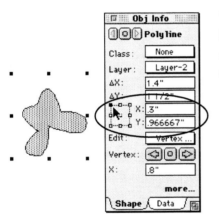

The object's stated position (x and y coordinates) are in fact those of the currently selected handle of the object. Click on a different handle in the palette and see how the position coordinates change.

This is the reason why we chose **Eight Selection Handles** *under* **Preferences...** *earlier: the default option of four handles produces less clutter on the drawing, but wouldn't illustrate the point as well, and would prevent you from resizing by one of the middle handles, which allows changing one dimension only*

As you can see, the **Obj Info** palette provides a convenient one-stop to various attributes of the object. The type of information varies according to the object selected. We will discuss the meaning of these attributes as we go along.

Obj Info *is an extension of the Data palette of previous versions, but also completely replaces the old* **Reshape...** *dialog and gives additional one-stop access to an object's Class and Layer as well.*

Drawing Walls

OK, time to do some drawing. Launch a new Untitled file from the Stationery we created earlier in the tutorial, with a view to produce a simple general arrangement plan.

Click on the **Wall Tool** (keyboard shortcut: 9).

Notice how, when you do so, a number of options appear in the Mode Bar:
    • The first three are about whether you are creating, joining, or 'cleaning up' a wall (in that order).

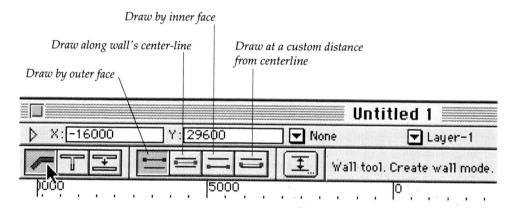

• The next four relate to how you are drawing the wall: by the inner face (assuming a clockwise motion), the outer face, its centerline or perhaps a certain distance offset from the centerline.

• The last item is a button which calls up the **Wall Preferences** dialog. This is where basic attributes of the wall are set, such as overall thickness (**Separation**), method of 'capping' (whether it ends with a flat line, a round one or none at all), whether it is one leaf or a composite or cavity wall, etc. These are

all hints at the fact that the Wall Tool does more than just draw a double line, but several operations all wrapped up in one (sometimes known as a **macro**). The result is an object which is a complex collection—or group—of lines and fills that works as one and can respond 'intelligently' to junctions with other walls and inserted items such as windows and doors. It is also a hybrid 2D/3D object, with an automatic Z height that can be set in advance for a given floor or storey (as we will see later).

For now:

1. Type **9** [@] or **225** [®] in the **Separation** field in the Wall Preferences dialog

> *No need to type the sign or abbreviation of the unit:*
>
> @: *Regardless of whether you chose* **Feet & Inches** *or* **Inches** *as your units, MiniCad assumes unsigned units are always inches: to indicate feet, you must type its sign (') afterwards*
>
> ®: *The program assumes unsigned units are whatever you've chosen under* **Unit...**: *mm, metre, cm, etc — in which case, indicate other measurements by decimal point. Thus, if your units are metres, type* **.225** *to indicate 225mm*

2. Click and hold on the **Caps** pop-up field, and select **Both** (this closes off wall sections at either end where left open)

> *Ignore* **Control Off** *for now – irrelevant for our purposes*

3• In the **Type** pop-up field, make sure it says **Flat** (Round caps don't join well at T junctions)
4• Click and hold on the central—'fill'— bit of the rectangle in the Attributes palette, and move the cursor to choose a suitable pattern, such as a standard diagonal.

> *This is to make the walls stand out graphically during our tutorial. Choosing a fill pattern when nothing is selected means that all surface type objects from now until further notice will have that pattern. This saves us having to select and give them this pattern retroactively. The thick line of the rectangle, by the way, governs the patterns of the line of any object (relatively rarely used, but useful sometimes).*

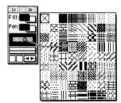

4. (®, *Optional*:) Click on the **Cavity Lines...** screenbutton. This brings up the **Cavity Setup** dialog (see box, next page).

With our wall configuration now ready, **OK** the Wall Preferences dialog, click on the upper lefthand side of the drawing area and start dragging the cursor to the right to start drawing.

Notice how, as you do, the Data Display Bar monitors your movements by analyzing them in terms of how far horizontally (ΔX) and vertically (ΔY) the cursor is in relation to the startpoint, the length (L) of the line so far, its A(ngle) (in relation to 0° = 3 o'clock), and the coordinates of the cursor at the moment.

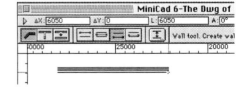

(continued on p.28)

On Making Cavity Walls

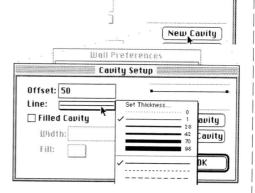

Click once on the **New Cavity** button...

A cavity line appears, with handles at each end indicating that it is selected, ready to be told what to be and where to go. In the **Offset** field type **50** (@: 2) Also, click and hold on the **Line** pop-up and note how you can choose a suitable line thickness (to distinguish it from the outer faces of the wall, which might be thicker).

Click anywhere in the white area of the dialog (don't press Enter: this confirms the setting but also dismisses the whole dialog) and note the line moves into position.

The measurements in the Cavity Setup are made in relation to the wall's centerline. This requires a little mental arithmetic: if, for the sake of argument, you want an outer leaf of 50mm (2") in a wall whose total thickness (Separation) is 250 (10"), you calculate that, in relation to the centerline, the inner face of the outer leaf must be offset 250 ÷ 2 - 50 = 75mm (10 ÷ 2 - 2 = 3").

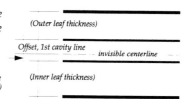

(Outer leaf thickness)

Offset, 1st cavity line   ⸻ invisible centerline ⸻

Offset, 2nd cavity line
(negative figure if below centerline)   (Inner leaf thickness)

Now click **New Cavity** again and type **-25** in the **Offset** field. (Being below the centerline, the figure is negative.) **OK** the dialog.

*Cavity components can optionally be fills instead of lines — to represent insulation, for example. This is triggered by turning on the **Fill Cavity** option, whereupon the **Width** and **Fill** fields are activated. The fill object is an open-ended rectangle (i.e. with only top and bottom faces drawn): type in the width and give it the fill of your choice from the pop-up. For its positioning, note that the offset here refers to its inner (bottom) face: to illustrate, note here how a 50mm (2") fill object relates to the previous two cavity lines when placed at zero offset.*

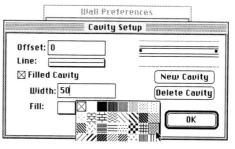

Incidentally, you are not limited to only four lines: walls can have any number of lines and/or fills. In addition, you can reselect a cavity line/fill post-creation and change its parameters at any time.

These fields are not just monitoring information, however. They also allow us to determine the size of the wall (or other object) as we draw, using numerical entry from the keyboard.

This is where the Click-drag Preference comes into play: with it on, when you release the cursor, the wall (or any other object) is set. With it off (as it should be, for our purposes), the wall will not be set until we click a second time. This gives us time to...

Press the `tab` key repeatedly: note how each field in the Bar gets highlighted in turn, eventually cycling back to ΔX. The highlighting means the field contents are selected, which in turn means that whatever you type at that moment will replace them, so that instead of reflecting the cursor's current position, the cursor now goes as you wish to. Any two of these fields is enough to provide the program with the information it needs to establish the line. What's more, this can — and is meant to — be done on the fly.

Try it. Having clicked on the lefthand side, press the `tab` key to highlight the ΔX field, and type (preferably using the numerical keypad—not the row at the top of the keyboard, for the sake of convenience) 10'3 [@] or 3130 [®], then press `Enter`

Note:

- *This is important—and it must be `Enter`, not `Return`: this confirms the input. Otherwise, the sizing doesn't 'take'.*
- *@: The foot sign (') is also critical: it tells MiniCad it's 14 feet, not inches*

Now play around with the cursor. Notice what happens: no matter where your cursor moves, it will only affect the wall's vertical extent. You are free to move the cursor up and down as much as like, but the horizontal extent of the wall (ΔX) is now set at 10'3" (®: 3128mm).

Note, too, that the ΔY is field is now selected (highlighted), ready for data entry. This means whatever you type now (and `Enter`) will be the wall's vertical extent and 'set' it. In this case, we can type 0 in the ΔY field and `Enter` to confirm the wall as a simple horizontal section.

As soon as you complete one section, the Wall Tool is ready for the next (if you only wanted a single section of wall, you need to double-click at this point), with its own ΔX, ΔY, etc. `Tab` straight to the ΔY field this time, type -2470 (@: -8'1) and `Enter`.

The minus is important: it tells the cursor to go down rather than up.

This time, however, instead of typing 0 in the ΔX field to indicate it should go straight downwards, press the Shift key as you drag the mouse: this constrains the wall to the nearest orthogonal, 45° or 30/60° angle.

Constrain to the vertical and click (once) to set the line.

As in most Mac draw applications, pressing the Shift *key constrains walls and lines to 30°, 45° and 90° increments— also rectangles into squares, ellipses into circles, etc.*

Notice how the sections—including cavities, if you have them—are joined up automatically at the corners as you draw

Make the next section—CD— 5770mm (18'-11") in the manner of AB.

For the diagonal section DE, use a different tack: here you may not know what the ΔX or ΔY extents of the section may be, but you might know its actual length and angle: tab to the L field, therefore, and type 2838 (9'4"), Enter, then tab to the A(ngle) field and type -45

Angles in MiniCad are negative when clockwise, with 0° being 3 o'clock. Lengths, on the other hand, are always positive, by definition.

Make section EF a vertical 8000 (26'3") long.

For sections FG and HK, remember to make the ΔX dimension negative, so that it goes left instead of right.

@: No need to type the dash in feet & inches dimensions (as displayed by the program), but for inch fractions, you must type a space before the fraction. Thus, for 12'-10 1/2, type 12'10(space)1/2

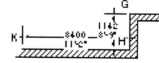

After point K, pause for a moment.

To find the point L that is aligned perpendicularly with startpoint G, there is no need to make any calculations or enter any numbers. Instead, we can take advantage of MiniCad's 'SmartCursor' technology.

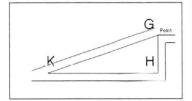

With Snap to Object on (press Q if it's not), drag the cursor from K and just touch point G, as if to say: 'You see this? I want to align with this point...'

Next, drag the cursor to the left. You needn't be too careful about it: now that it knows which point you're referring to, the SmartCursor will indicate, with a dotted line, whenever you are precisely aligned with it (Align V = Aligned Vertically)

When you are roughly perpendicular with K, press Shift to constrain it precisely to the vertical, look for confirmation of the Intersection point between it and the alignment with G and click to set.

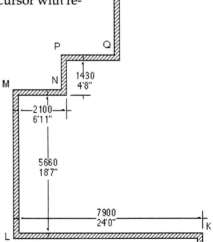

Carry on in this manner, entering the remainder of the building outline to the dimensions shown. Find point Q using the SmartCursor with regards to point A as we did with point L.

When you bring the cursor up and snap to A. It may or may not say Point, but will definitely have an indicator spot to the lower right). Double-click to set.

A mitered corner should form automatically between this last section and the first: if it doesn't, join them manually using the Joining Walls mode in the Mode Bar. The result should look like this:

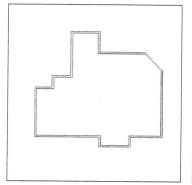
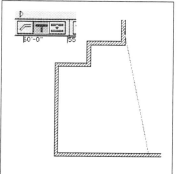
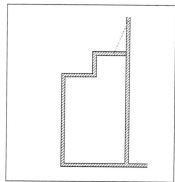

We're almost there.

Click on the Join Mode button, then click, drag and click between the last section and section LM. See it join in a cleaned-up T-junction.

This makes the old junction at Q look like a butt join now. So repeat the operation between the two walls there, to clean it up.

In case you're wondering, the lower left space is to be a garage and 'mudroom', hence the extension of the external wall within the footprint.

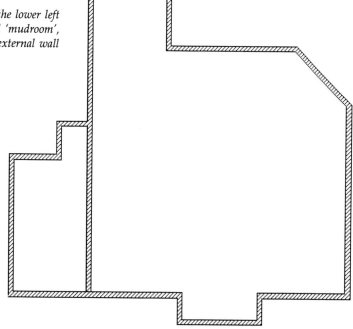

Congratulations. We have completed the outline of the external walls of the house.

Save the file, and have a break.

Now for some internal walls.

 Select (click once on) the **Zoom In Tool,**
then click-drag-and-click an imaginary rectangle…

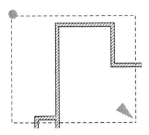

> *A 'marquee' in the jargon—so-called because of the running dotted line, reminiscent of Broadway theater displays*

…around the top left part of the building plan.

 Choose the Wall Tool again, and remember to click again on the Create Wall mode button as, having left it in Join Wall mode, it did *not* default back to Create Wall, as one might expect. Then go into the Wall Preferences dialog again and change the Separation to **120** (@: **4.75**).

Note:
> @: *This demonstrates that you can enter digital data even when working in fractional Feet & Inches Units. MiniCad then converts it into fractions for you*
>
> ®: *If you had cavity lines in the previous settings, go into the Cavity Lines dialog and click repeatedly on the* **Delete Cavity** *button until they are all gone*

Now…

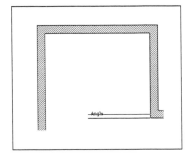

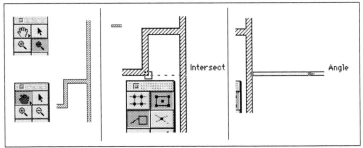

With Snap to Object on, click to and drag a wall from the inside corner at **C**, press Command or I key until it is drawing from bottom face as shown, then press Shift to constrain it to the horizontal.

Click to set halfway.

Double-click on the **Zoom Out Tool** (or double-press the v key), then choose the **Pan Tool** (press z). Click and drag the drawing upwards—as if it were a piece of paper on the drawing board—to bring the mud-room into view. With Snap to Object on, choose the Wall Tool again, touch the corner of the mudroom with the garage, & move it in Alignment to the right till you reach the inside face of the long wall. Now press the s key on its own: this turns on Snap to Surface. When the screenhint Intersect confirms, shift-drag out a wall flush with the corner.

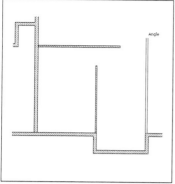

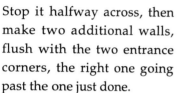

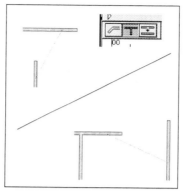

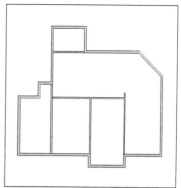

Stop it halfway across, then make two additional walls, flush with the two entrance corners, the right one going past the one just done.

Join up the three in T-junctions as shown, and while you're at it, join the very first one we made earlier the external wall opposite it.

The result should look like this.

When a Butt-Join is Called For

The junctions made by the Join command (or its keyboard shortcut, `Command-J`), are usually 'cleaned up' automatically.

I say 'usually' because, as you may have noticed, MiniCad refuses to do so at when joining a wall to an existing L-junction, such as the one at C or the two entrance ones. In those cases, it is best to keep them as found. (The Y-Join Tool—to the right of the button Join Tool below— is supposed to solve this problem but doesn't)

You may not want a cleaned-up junction, though, but a butt-join.

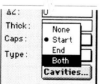

To do this retroactively, select the wall in question (being a Group, one click anywhere on it suffices), open its Object Info palette (`Command-I`), and choose Both from the Caps pop-up menu.

To get the right result from the outset, instead of the Wall Tool/Join Mode choose the special **Butt-Join Tool**—one of the pop-out alternatives to the Wall Tool. Apply in the same manner as the ordinary Join Mode.

We now come to one of the most fundamental and useful features of MiniCad drafting—one which nevertheless often goes overlooked by newcomers and is not given anything like its due in the program manuals. This is the **Floating Datum**, and it is best introduced by example—in this case, the creation of internal walls that are not necessarily situated at convenient points on the external envelope, but at specific distances of our choosing.

With Snap to Object on, and the Wall Tool in Create Wall mode…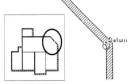

…Bring the cursor to 'touch' the inside of corner E and—without click-ing—pull away slightly down and to the right until the word Datum ap-pears, with a circle centered on the corner.

 Note:

Pulling down and to the right usually works for me, but on occasion you may need to move it about somewhere in the vicinity of the point in question, with all sorts of other screen hints appearing instead.

Don't settle for anything else, though, and if it refuses to appear, try zooming in and trying again, using the Zoom In Tool as we did earlier.

Now look at the Mode Bar: note that it now says X and Y. These are coordinate references—in contrast to the ΔX, ΔY we saw earlier, which are dimensions. This is because point E is now considered a temporary datum, or '0:0' point. The figures we type in now are therefore coordi-nates of the point from which our wall will *start*.

Type -1700 (@: -5'7) in the Y: field, and Enter.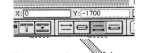

Notice our friend the dotted line marking that distance down from point E. Instead of typing 0 in the x field, bring the cursor to intersection of that line with the inside of the wall. When it confirms with the screen hint Intersect, click and drag to the left.

Congratulations. In one operation, you've started a new wall at a distance of -1700mm (5'-7") from a given corner. Press Shift to constrain the wall to the horizontal and double-click to set it anywhere short of the last vertical wall we made.

Any snappable point can be nominated as a Floating Datum for the creation or placing of objects at a distance defined by you. This makes it an extraordinarily useful and timesaving tool.

By the same method, make another wall parallel to this one, using its bottom inside corner as the new Datum (touch without clicking, then move away and `tab`...), whose top face is -1900mm (-6'-3") below it (Y field). Constrain to the horizontal, and make it about the same length as the last one.

Keep going in this way, making additional internal walls (shown here in darker pattern) at the offsets shown. When making the section emanating from point **C'**, make sure to make it 3680mm (12'-1") long (Y field).

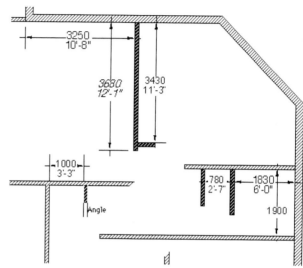

Then join up the various sections as shown.

The order doesn't matter, but to avoid having to go over a T-junction to clean it up, it's generally best to do the L-junctions first.

Then save the file (`Command-S`).

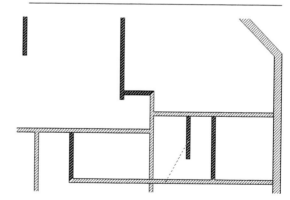

§

Dimensioning

We have been pre-sizing walls and placing them at specific distances all over the place, taking it on trust that these sizes and distances are indeed what we asked them to be. Now is a good time to test whether this is in fact the case, using the dimensioning tools.

Unlike previous versions, the dimensioning tools are now on a palette separate from the 2D tools. This is not opened along with the others, so we need to call it up via the **Windows** menu, submenu **Tools**.

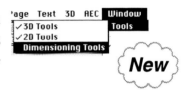

And this is what it looks like. Their icons are fairly self-explanatory, but we will learn each in turn as we go along. With the exception of the bottom two, they all have keyboard shortcuts, which in practice are far more convenient than pulling down the palette every time.

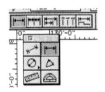

Press the N key: as the palette shows, this chooses what is called the **Constrained Dimension Tool**, which measures the horizontal and vertical extents of objects—as distinct from the **Unconstrained Dimension Tool** next to it (hotkey: M) which can measure diagonal distances.

There are five different modes to choose from. The first one—and default (indicated in the illustration)—is what Graphsoft call the **Constrained Linear** and I think of as simply Point-to Point mode.

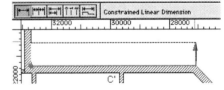

Try it on the external corners of the wall section **CD** (as ever, with Snap to Object on): click at **C**, then at **D**, then pull up to where you would like the dimension to be, and click to set.

The result will be displayed—in the metric or English units of your choice earlier.

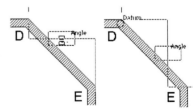

Now try it on the diagonal section **DE**: notice how, prior to setting, the 'ghosted' dimension refuses to measure the actual diagonal, but jumps from a horizontal to vertical measurement of the section.

Click to set either one, then choose the Selection Tool (hotkey: x), click once on the dimension that formed, and press delete on the keyboard to remove it.

- *That, by the way—in case you didn't know—is how you delete objects.*
- *Like walls, dimensions are pre-grouped collections of their components: a single click selects all.*

 Now try the Unconstrained Dimension Tool (or Diagonal Dimension Tool, as I prefer to call it). Choose it from the palette—or press M—then dimension DE in the same manner, using the default Linear mode as the other one.

Compare and contrast the different result:

For your next trick, go back to the Constrained Dimension Tool and try its third Mode— **Chain Dimensioning**—applying it to the bottom range of external corners of the building.

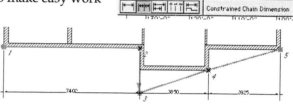

As its name suggests, this was conceived to make easy work of producing a neat consecutive chains of dimensions. It works just like the Linear mode, only after producing the first dimension, you only need to click on the next points to be measured to for the dimensions to be produced automatically on the same line as the first (i.e., no need to pull out). The tricky bit is to know how to stop: *it will continue like this indefinitely until and unless you double-click* (in this case, at clickpoint #5).

Baseline Dimensioning works in exactly the same way, but the result is different.

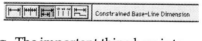

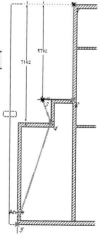

Try it along the lefthand side of the building. The important thing here is to allow enough room, taking into account the step-like progression of the offsets.

Chain and Baseline modes are available in Unconstrained ('Diagonal') Dimensioning as well. Constrained Dimensioning, as you can see, has two additional modes which we shall acquaint ourselves with later. Meanwhile, practise applying these various options throughout the rest of the plan.

Classes

We now have a drawing fairly cluttered with dimensions that will get in the way of any detail. How to remove them without actually deleting them?

In traditional CAD programs one might assign temporarily unwanted objects to a layer that was set to be invisible. MiniCad's concept of **Classes** achieves the same objective more elegantly and flexibly. These are logical and user-defined categories of objects which transcend all layers and can be singled out at any time for the purpose of display or analysis. The dimension clutter provides an excellent opportunity to see how it works.

Choose **Classes...** from the **Organize** menu.

In the dialog that follows, note that two Classes already exist by default: **Dimension**, and **None**. As they suggest, the program makes no assumptions as to any other we might like to make, but it does conveniently and automatically assign all dimensions to a discrete Class of their own, while all other objects are—pending your own choices—placed in **None**.

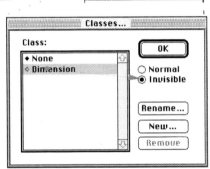

The black diamonds to the left of the Class names mean that they are both currently **Normal**, i.e. visible from within other Classes. Select the **Dimension** class and click on the **Invisible** radio button on the right. The diamond goes hollow, and—after **OK**'ing the dialog—the actual dimensions should disappear, as well, leaving the building walls pristine once more.

> *This is because we are currently 'sitting in' the None class (a default setting). If we were sitting in the Dimension class—i.e., if it were 'active'— its members would be visible. Which is just as well, as otherwise there would be no way to view or edit members of invisible Classes once they had been made so.*
> *Incidentally, a Class is made active by selecting it when no object is selected. To ensure this, double-click on a blank part of the drawing—or better still on the Selection Tool.*

Before we proceed any further with the drawing, call up the **Classes...** dialog again, only this time—and on all future occasions, since it's easier), do so via the pop-down menu in the Data Display Bar currently labelled None.

 The label here tells us which is the current active Class when nothing is selected.
When an object is selected, its Class is displayed instead.
When two or more items are selected that don't belong to the same Class, the labels says ?

Now click on the button **New...** button, and—in the ensuing dialog—type **External Walls**, then Enter or Return to confirm the entry and return to the Classes... dialog. Click again on **New...** and this time type **Internal Walls**, and Return or Enter. Then Return again to **OK** the Classes... dialog and return to the drawing.

We will now assign objects to the new classes.

While pressing Shift, click on all the external walls in turn to select them.

> *Pressing* Shift *allows us to make multiple selections by consecutive clicks on object: without it each selection would deselect the previous ones.*

Without clicking anywhere else (which would deselect them), go to the Class pop-down menu as before, choose **External Walls**...—and release.

That's it. That's all you need to do to assign an object to a given Class.

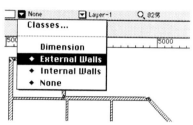

To do the same thing for the Internal Walls, use a handy MiniCad shortcut for multiple selections. Still with the Selection Tool, press and hold the Option key, then draw a marquee that encompasses or touches all the internal walls as shown.

> *Without the* Option *key, only objects that were fully encompassed by the marquee would be selected. That is the feature's plus—but also its drawback, since you can easily select objects you don't intend to. If that does happen, press* Shift *immediately afterwards and click on the unintended items: this deselects them individually without deselecting the others.*

Then release and—while they're still selected—release Option, press Shift and click on the one we missed in the upper lefthand corner to add it to the collection.

Then assign them to the Internal Walls class by choosing it from the Class pop-down menu.

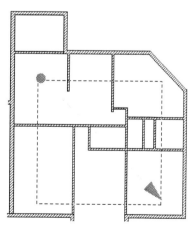

Before we proceed, a brief exercise to apply a few of the things we have learned and one or two others which should come in handy in your work.

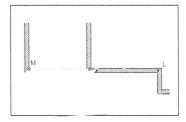

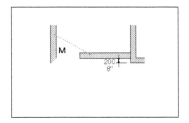

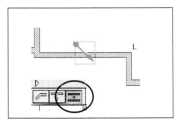

Click to select wall section LM, then click-drag its left handle back beyond the right wall of the garage.

Unlike ordinary groups, walls have only two handles—one at each end—for resizing purposes. This theme recurs in their 3D editing.

Using the Floating Datum, and Wall Tool, make a new wall section offset from the new corner at 200mm (8") and join it to the existing walls at both ends.

Not the most elegant of arrangements, but it's the principle that counts.

Notice the line removed in the area of the old junction. This is repaired by drawing a marquee around it with the Wall Tool in Remove wall breaks/caps mode.

Apart from the cosmetic repair of redundant junctions, this mode is also often used to 'disconnect' walls that have been joined together.

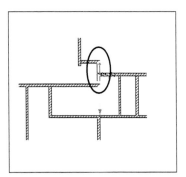

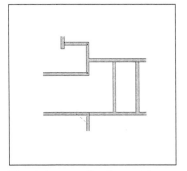

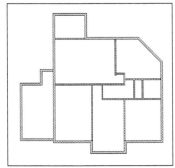

Do the same for the vertical internal wall flush with the right side of the entrance, and make a new short section, as shown.

The join up all relevant junctions.

(The original wall helped to establish junction points, hence it wasn't drawn like this to begin with)

The final plan should look something like this.

Save your file.

Text & Other Tools

Text

To help orientate ourselves in the plan, and as an opportunity to use some other tools, let us now label the spaces.

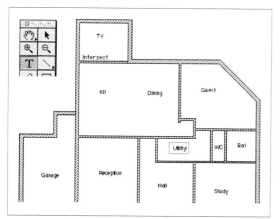

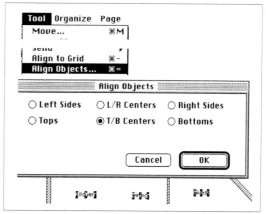

Choose the **Text Tool** (hotkey: 1), and click in the middle of each space in turn to type its name as above. Don't worry about their precise placement just now...

...because once completed, you can align them neatly by selecting the ones you want, then calling up **Align Objects...** from the **Tool** menu (shortcut: Command=), and choosing **T/B** (='Top/ Bottom') **Centers,** & **OK**

Alternatively, you could align their Bottoms, as it were, in which case all objects would take their cue from whichever was bottom-most. Or Tops, to take the cue from the one that was topmost. That's the principle. Note that each bit of text here is a separate object (a 'textblock') with its own handles. A textblock can, however, be many paragraphs long.

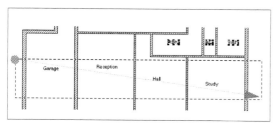

An ordinary marquee with the Selection Tool (no Option key this time) can speed up the selection of several objects at once.

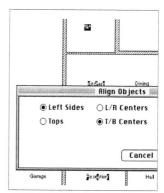

The labels can be aligned by their **Left Sides,** too: but make sure in this instance to click *again* on the **T/B Centers** to turn it off.

Through the **Text** menu, you have the usual access to whichever fonts are installed in your system, text sizes and styles—as well as additional options (more relevant to larger sections of text) such as line spacing, justification etc.

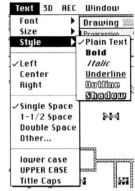

> *Of these, what I find rather useful is the ability to change between lower-case, Title Case (where the first letter is capitalized), or all upper-case. Unlike most other programs, these are not mere temporary display formatting, but proper transformations of the letters concerned from one to the other—a feature not available even in most wordprocessors, and one which can come in handy sometimes.*

To ensure that you have your preferred font, size, etc. each time you type text, choose your options before selecting the Text Tool. This makes them the default, which can also be saved as part of a Stationery file.

Room Areas

You and I may be able to look at an area bounded by several walls and see it as a space, but in software that level and type of intelligence is still not generally available in CAD software. For MiniCad to know which space is which, we must spell it out by defining the rooms as actual geometric objects, i.e. rectangles or polygons.

One way to do this would be to choose the **Rectangle** or **Polygon Tool**, as appropriate

> *Don't ask me what the hotkey for that is—a guy has to draw the line somewhere*

and, with Snap to Object on, create polygons whose vertices snap to the appropriate corners. With rectangular rooms it's not so bad; with others it can be good practice of the Polygon Tool, but there's an easier way. This is to take advantage of MiniCad's **Combine to Surface** feature and the fact that walls can each be selected with one click only.

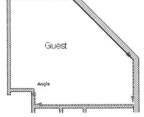

Before you do, with nothing selected in the drawing (double-click on the Selection Tool or double-press x), choose a light, general dotted pattern from the Attributes palette. This makes it the default pattern for all new surface-type objects until further notice, so the room areas created will immediately be visible.

Zoom in on a given room that is fully enclosed by walls that have been

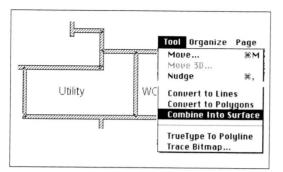

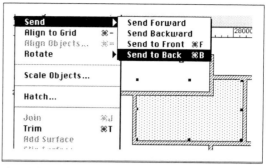

properly joined

Didn't join properly? This is where you get found out.

Shift-click on all its bounding walls, then choose **Combine into Surface** from the **Tool** menu. The cursor turns into a bucket icon: click once with it in the middle of the room. A polygon corresponding to the enclosed area

should form, with the fill pattern just chosen.

Because it was created more recently, the polygon now conceals the text label that was there. Click on it to select it, and **Send** it **to** the **Back** via the **Tool** menu or by pressing Command-B

For Combine into Surface to work, it is essential that:

 Note: *a] The walls have been properly joined or overlap, i.e., the space doesn't 'leak' anywhere, and*
b] You haven't selected any objects <u>other</u> than the walls, lines or other objects bounding the area

Even then, however, it can be a temperamental tool, requiring two or three tries on occasion. This is particularly true of areas with many sides and/or internal angles greater than 180° (e.g., the Guest room here).

In these situations, there may be nothing for it but for you to draw the polygon manually. However, for this, too, there are shortcuts: instead of snapping to the corners with the Polygon Tool you can compose the shape out of several overlapping rectangles, which you then add together to make one shape: Shift-*select them, then choose* **Add Surface** *from the* **Tool** *menu (no keyboard shortcut). A handy feature for all sorts of situations.*

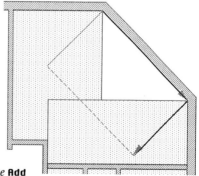

On other occasions, in an effort to get it right the program makes a stab at a shape which includes the walls themselves (see right), resulting in their changing fill: in these situations you should immediately **Undo** *the result (* **Edit** *menu—or* Command-Z *on your keyboard: a shortcut well worth remembering; but it only works immediately after the wrong move).*

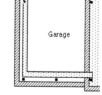

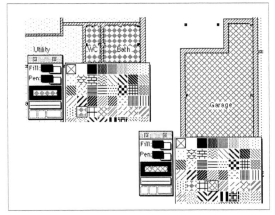

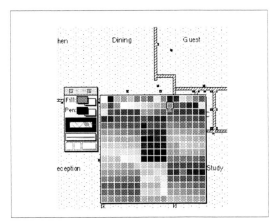

After repeating the process for all internal spaces, select the WC and Bathroom room areas and give them a suitable tile pattern from the Attributes palette.

Choose yet another pattern for the Garage.

This is both for presentation purposes and to enable automatic analysis by the program later

To make all these fill patterns less obtrusive on screen, select all the room areas and choose a light gray for the pixels that are currently black (known as the **Fill Foreground** color), by click-and-holding on the little foreground rectangle next to Fill in the Attributes palette

These will also print lighter in grayscale or color printing.

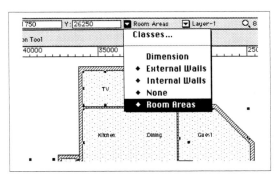

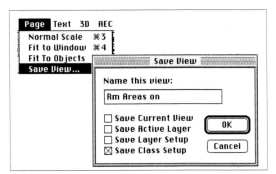

Make a new Class for the Room Areas and batch-assign the polygons to it.

Before making this Class invisible, take advantage of another MiniCad feature to make it easier to return to the visible setting later:

Choose **Save View... (Page** menu). In the dialog that follows, turn off the first two (default) checkboxes, and turn on **Save Class Setup** only. In the text box at the top, type something like **Rm Areas On**, then Return or Enter to **OK** the dialog...

Congratulations. You have made your first custom Command.

Preserving the particular Class Setup—as we have just done—means that the particular combination of Normal and Invisible Classes that existed when the Command was created can be returned to at any time in future by simply double-clicking on it. This is far more convenient than recreating it manually through the Class Setup dialog.

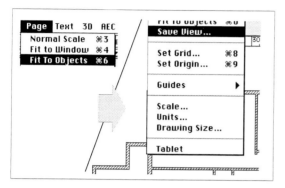

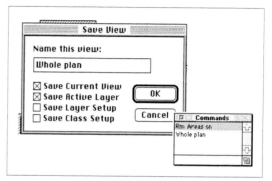

Since we're making Commands, let's make another one of a simpler kind. Choose **Fit to Objects** (**Page** menu, or `Command-6`) so that the entire plan just fits inside the window. Then **Save View...** again—

...and this time keep the default settings. Name it **Whole Plan** or some such, then **OK**. The resulting Command is automatically added to the palette, and its job will be to return us in future to this particular View and Layer of the drawing from anywhere else we happen to be.

 Note: *By saving the Current View, the program is referring only to this part of the drawing sheet: if you move the actual objects somewhere else you have to update or redo the Command manually.*

In preparation for the next bit, change the default fill pattern back to white.

Save your file.

§

Symbols

General

As objects that repeat themselves throughout a drawing, doors and windows are classic examples of MiniCad's **Symbol** facility, which is expressly designed to make the duplication and editing of repeated graphic objects easier and more economical with system memory

> *That's **RAM**—Random Access Memory, akin to remembering to 'carry 1' or whatever when adding numbers—not to be confused with **hard disk space**, which is long-term memory, if you like. Although that, too, benefits, since using symbols results in smaller file sizes*

Symbols are, in fact, duplicates of a special kind: they can be placed on the drawing with a single click, and are universally linked to their original prototype in such a way that changes made to any one are reflected in all the others as well. They have many other advantages, which we shall appreciate as go along and which will hopefully persuade you of their preference over simple duplicates.

The strategy is to regard as a prototype the first example of any item that is likely to be repeated more than twice or three times in the drawing *or to recur regularly in other drawings*. This then is stored away in the file's **Resources** and acts as the master symbol, of which **instances** are then be placed in the drawing, as often as we like (memory allowing). Symbols may be 2D only, or 'hybrid' 2D/3D, which means they have one appearance in 2D drawings, and another in the 3D model of the building (in the absence of a specially-designed 2D aspect, 3D-only symbols appear in plan as they would in Top View). In the Tutorial Manual that came with the program, the 3D part of a symbol is made first, but since we are approaching the subject from a drafting point of view, we will do the opposite.

Incidentally, doors & windows are slightly special cases of symbols, as they are placed in walls which, as hybrid 2D/3D entities, place certain restrictions on the handling of such objects (which shall become apparent in due course). However, we will start with them as they are logically the next step to do in our drawing, and because the others will then seem that much easier.

Windows

In drawing the prototype/master of a window, it is not essential but it is easier if we do so in the context of the external wall for which it is intended.

In the best tradition of North American drafting, windows are represented in general arrangement drawings as plain openings in the wall, with details such as projecting sills etc. left for depictions of higher detail.

This explains MiniCad's approach in the creation of window symbols. In drawing one at this scale, you should focus on creating a simple rectangle of the same depth as its intended wall, as wide as the structural opening, and (typically) simple small rectangles on either side representing the jambs and a single line for the glazing.

Being flush with both faces of the wall is particularly important, as it means that the center of the window's depth then corresponds to the centerline of the wall, which is where MiniCad insists on snapping the placement point of any a symbol placed in it.

Trained as I am in a different tradition, I admit I resisted this approach throughout all previous versions of the program, in the hope of seeing this particular feature change. This hasn't happened yet (although it is in the works—see Note, p.50) but life is too short, and I have succumbed.

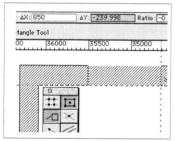 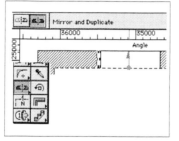 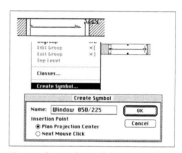

Zoom onto the top wall of the TV room and—with Snap to Object & Snap to Surface on—choose the Rectangle Tool and make one that snaps between the two surfaces of the wall, `tab`-&-typing a width (ΔX) of 850 (2'9), & `Enter`.

By the same method, make a left jamb of 50 (2") from the upper-lefthand corner of the same rectangle. Then choose the Mirror Tool—Mirror & Duplicate mode—and draw an imaginary reflection axis from the Bottom Center to the Top Center of the rectangle (or vice versa), and release. This should create the right jamb in the correct position.

Complete the object with a single line from Center Right of one jamb to Center Left of the other, then draw a marquee over all of it with the Selection Tool to select everything, and choose **Create Symbol... (Organize)**.

Name it **Window 850/225** (@:**Window 2'9"/9"**), and **OK** the dialog with its default Insertion Point.

Well done—you've made your first Symbol. The prototype window has disappeared from the drawing because it is now stored in the file's **Symbol Library** (i.e. part of its Resources).

 To place an instance of it in the drawing, we double-click the **Symbol Insert Tool**.

This calls up the **Resources Palette**: the one-stop shop for symbols, Commands, hatch-types, worksheets, Records etc. of this and other MiniCad files on all volumes accessible from your hard disk.

So far we have been acquainted with Symbols and Commands. We will cover the other types of Resources in due course.

Note that our window symbol appears in the list box, under a pop-down menu displaying the name of our current file. A depiction of the symbol in miniature appears in the righthand box near the bottom: this is to remind us what it looks like, and applies to whichever symbol is currently selected in the list (having just been made, our window is it). The box on the lower left displays whichever is the current *active* symbol, and it is initially empty. Click on the '«' button to make our Window the new active symbol. Now you are ready to place

Since the Resources palette is a palette, not a dialog, there is no need to dismiss it first

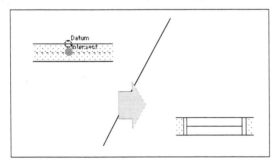

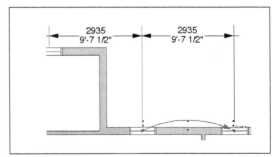

With Snap to Object and Snap to Surface on, bring your cursor over to the middle of the wall section we created the prototype in until you get confirmation of its Top Center. Then pull down in Alignment with it till the Intersect point with the centerline of the wall.

Click once—and the window is placed.

The Symbol Insert Tool carries on placing until you choose another tool. With single clicks and using the Floating Datum (yes, it works for this, too), place another two instances along the Dining and Guest walls at equal offsets of 2935 mm (9'-7 1/2") from the first.

Pan the drawing (hotkey: z) a little up and to

the right, to reveal the left side wall of the TV Room and Kitchen, then choose the Symbol Insert Tool again and click once more—aided by the Floating Datum—a distance (ΔY) of -3670 (-12'0") from the top (outside) corner.

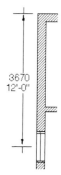

3670
12'-0"

> *Two things to notice here:*
>
> *a] The symbol automatically takes on the orientation of the wall, although it was created at a different one. This is one of the advantages of true symbols in true walls as opposed to ordinary duplicates, and it is particularly useful where the angle of orientation is unknown or complicated (outside walls, of course, the original orientation of a symbol applies)*
>
> *b] Although you changed tools since the last placement, the program still remembers which is your active symbol. This is true no matter how many tools you use in the interim, until the file is actually closed.*

For our next trick, we shall demonstrate how a new symbol is made out of an existing one. Quick-click the Symbol Insert Tool in a empty part of the drawing (anywhere other on or near a

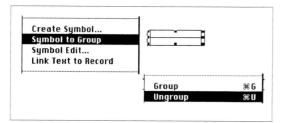

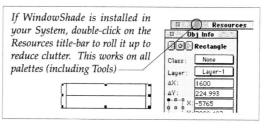

wall) to create another instance of our window. While it is still selected, choose **Symbol to Group** (**Organize** menu).

> *This 'de-symbolizes' the symbol, turning it into an ordinary Group of objects. This allows us to make changes to this object without affecting the existing instances of our first window.*

Then **Ungroup** it (same menu—or Command-U as indicated). Click in a blank part of the draw-

ing to deselect the components, then click on the main rectangle (structural opening). Press Command-I to call up its Object Info palette, check that its point of reference is one of its left handles, and type 1600 (@:5'3") as its new ΔX value. Press Enter to apply.

> *In case you're wondering, we couldn't just resize the whole Group, as that would have reduced the width of the jambs, too.*

Note: *We haven't discussed the concept of Groups in any detail as yet, but they're fairly self-explanatory, the idea being that several objects can be grouped (**Organize** menu, or Command-G) as and when necessary for the advantages of behaving as one object. Symbols are treated as special groups as they usually comprise several objects each. We will return to the subject further on.*

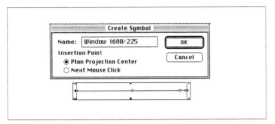

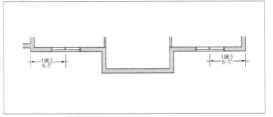

Drag the right handle of the glazing line to resize it to the new dimension, then marquee-select all of it, and **Create Symbol...** to name it **Window 1600/225** (@:**Window 5'-3/9**), keeping its insertion point at its center in Plan, as per the default

Since we have just made it, that is now our active symbol, so there is no need to specify it from the Resources palette: click just once on the Symbol Insert Tool, and use the Floating Datum to place two instances of the new window along the front wall, each at a distance of 1963 (6'5") from its outside corner.

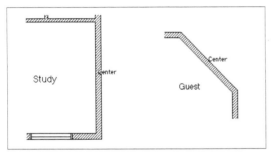

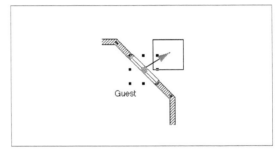

Place another instance at the Center of the other wall of the Study, and yet another at the Center of the diagonal Guest wall.

> *The program knows where the center of the wall section is because we are using proper walls that have been correctly joined. Otherwise it doesn't.*

Now, in preparation for its conversion into a French double-door, click-and-drag the new Guest window out of the wall. Note its resistance, and that the moment you click on it that whole wall section gets selected, too.

 Note:

One of the most pronounced idiosyncrasies of walls is that they behave very 'maternally' towards symbols placed in them: as its selection handles suggest, if you were simply to delete the symbol, the wall section would be deleted, too, in sympathy. If you do this by mistake (as often happens), Command-Z *immediately to Undo the operation and bring them back.*

Incidentally, as you click a symbol in place you may notice a brief appearance of four modes for 2D Symbol Insertion. This is an undocumented feature, conceived to let you to decide, on the fly, the point by which you are inserting the symbol. In fact, at the time of writing it is still under development and not yet operational. For now, ignore it.

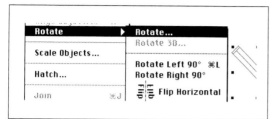

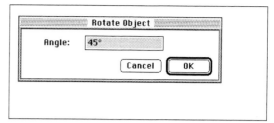

Symbol to Group the window instance to de-symbolize it, and—to simplify its editing—make it horizontal again by choosing **Rotate** (**Tool** menu), submenu **Rotate...**, and release

Type **45** and `Return` or `Enter` to confirm

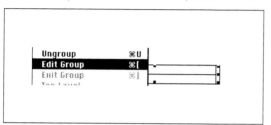

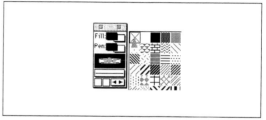

The window is now an ordinary Group. This time, instead of Ungrouping it, **Edit Group**

*Like keyhole surgery, **Edit Group** (and its cousin, **Edit Symbol...**) allows you to mess around inside a group without opening it all up first. Notice how the rest of the drawing temporarily disappears as a result: it's just you and the Group's components.*

Before we create the door swings, click somewhere blank to deselect everything, and change the default fill pattern to 'None' by click-and-holding inside the thick-line rectangle of the Attributes palette and dragging the cursor over to the checkbox-like square.

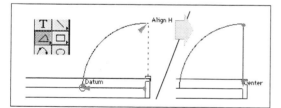

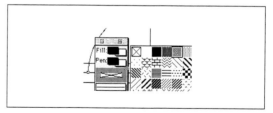

With the **Radial Arc Tool** in default Mode (radius-then-arc), click on the Center Left of the right jamb, then at the center of the glazing line, and drag the arc clockwise until it is Aligned H(orizontally) with the prospective hinge point. Click to set.

Next (this bit is entirely optional), with the arc still selected, click-and-hold on the *frame* of the thick-line rectangle in the Attributes palette and choose a uniform gray pattern

This patterns the line of an object, instead of its fill. A handy & more versatile alternative to dotted lines.

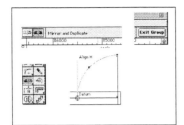

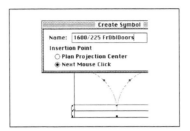

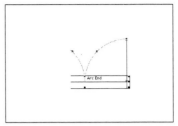

Use the **Single Line Tool** to draw the door leaf from the Arc's end to the hinge point, then select both leaf & swing and—as earlier—Mirror-Duplicate them to create the left door. Then click on the button above right to **Exit Group**

Since it is a group, one click on the collection is enough to select it. Then call up **Create Symbol...**, name it something appropriate, only this time click the **Next Mouse Click** option for its future Insertion Point. Then **OK**.

The cursor changes to a bull's-eye, to help you aim with your next mouse click at the object's future insertion point as a symbol. Click at the Center of the structural opening (the Arcs' endpoint).

With the Symbol Insert Tool, place the new symbol at the center of the Guest room diagonal wall—only, before releasing the mouse, move the cursor alternately up to the right and down to the left. Note how a ghost image of the symbol flips between door-swings-in and door-swings-out. Bring the cursor up and to the right to make it door-swings-out, and click to set.

> *This is another attribute of true walls, and its principal advantage is it dispenses with the need to rotate the symbol after placement or—as we shall see presently—to create right-handed and left-handed versions of the same door.*

Doors

The description of French doors leads us neatly to the issue of doors proper. Here, the procedure is much the same, with structural opening rectangles, radial arcs for door swings, a single line for the leaf etc. is much the same. The main difference is that structural rectangle is made with an invisible boundary line, to create the illusion of a cut in the wall.

> *Strictly speaking, the rectangle isn't necessary, as the program ensures a 'hole' is made in the wall to accommodate the door once it is inserted. However, it helps in the door's construction, and it will be handy should you decide to **Ungroup** your walls in order to be free of the idiosyncrasies of 'true' walls (which, incidentally, I don't recommend, as on balance 'true' walls are better).*

As with windows, it is best to create them in the context of their intended walls. But first, change your default fill back white again.

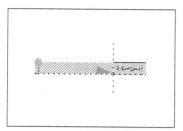

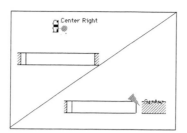

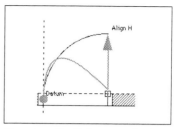

Zoom in on a typical—preferably horizontal— bit of internal wall for this purpose. Create a rectangle from Surface to Surface of a width 760mm (2'6"). Create a left jamb of 50 (2") as before.

This time, however, instead of Mirror-Duplicating it, do a simple Duplicate (Command-D). Then bring the cursor close to (but not on!) the duplicate's Center Right handle, and drag it to snap to the Center Right of the structural opening

Create a Radial arc as before from the Center Right of the left jamb.

If the Align H *doesn't indicate, bring the cursor momentarily to touch the corner of the jamb and pull up again*

Note: *The cursor in MiniCad changes into umpteen different shapes that change according to the current operation. None is more common or useful, perhaps, than the 'Snap-drag' cursor, recognizable by its distinctive 'crusader cross' look. It is your indication that MiniCad understands that you wish to move the selected object(s) by the nearest handle, possibly with a view to snap that handle to some other point. It is activated, as we've just seen, by bringing the cursor near—but not on—said handle (directly on it results in the Resize cursor—a diagonal with opposing arrowheads)*

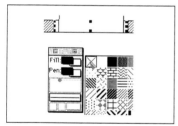

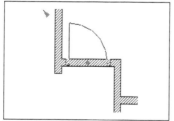

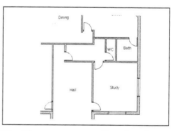

Select the structural opening rectangle, and give its line a 'None' pattern to make it invisible. Marquee-select all, **Create Symbol...**, name it, make its Insertion point the Center of the rectangle...

...and place an instance of it at the entrance to the Guest room, moving the cursor up and to the left between click-and-release to make the door-swing as shown.

By the same method, place additional instances at all these places.

Use the Floating Datum, or eyeball it, whichever you prefer. If you don't get it quite right you can always slide the symbol along the wall to get it right.

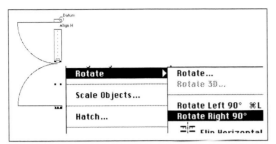

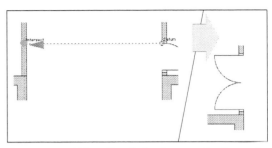

Pull out the entrance door to the Reception room, **Symbol to Group** it and turn it into a double door as we did earlier with the French ones (not forgetting to resize the invisible structural opening rectangle, too!). Only make sure you **Rotate** the finished product **Right 90°** before turning it into a Symbol: for some reason the program discriminates against wall-based symbols that have been created vertically.

As soon as it is a Symbol, place an instance in the Reception wall, after touching the inside of the Study door top jamb to Align with it during placement.

Afterwards place another one at the center of the wall between Reception and Kitchen.

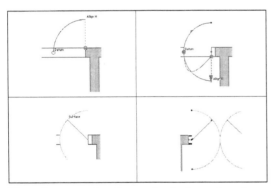

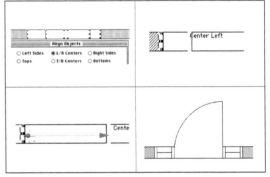

Variations on the theme include swing double-doors to the kitchen: in a structural opening 850mm (2'-8") wide, the radial arc, once made, is pulled with the Selection Tool into 180° (aligning with the jamb as shown). The door leaf snaps to the Surface of the arc, and the whole lot is then mirror-duplicated, made into a Symbol, and re-inserted at this wall section's Center.

The front door might be made from two overlapping rectangles of 1800 & 1100 (5'11" & 3'7") respectively, which are then aligned by their L/R Centers (**Align Objects**: Command =). Here, duplicate the first jamb by pressing Option before click-and-dragging it, the Snap-drag cursor ensuring correct placement as you do.

> Option-*click-drag is by far the most convenient way to create ordinary duplicates*

Furnishings & Fittings

These provide us with opportunities to employ other tools:

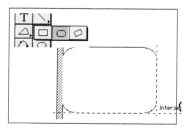

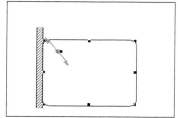

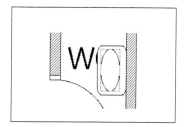

A rounded rectangle (tab-and-type dimensions) for a double bed…

…its 'roundedness' change-able by hand by means of the fifth handle (or through its Object Info palette)

Another such rectangle with an ellipse makes a basin…

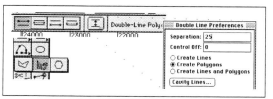

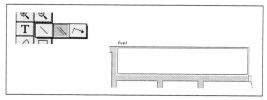

A double-line polygon of a suitable setting…

Note how its Modes & settings are similar to those of walls

can be used to create a cupboard (wardrobe), with ordinary double-line sections providing partitions

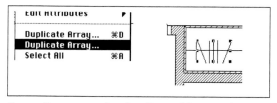

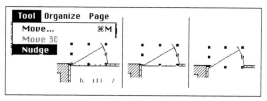

Some items may be duplicated in a series, using **Duplicate Array...** (**Edit** menu—or Option-Command-D)

or **Nudge**d gently (Command-, [comma]) into place, one pixel at a time, using the arrow keys

The more zoomed-in you are, the smaller this distance is in 'real' terms

Other fittings may be made of rectangles **Add Sur-face**d together, Mirror-Duplicated to the other side of the room then **Clip**ped by another **Surface** object

Whichever object is in front clips the one behind

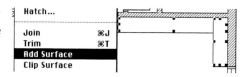

Duplicating by Dialog...

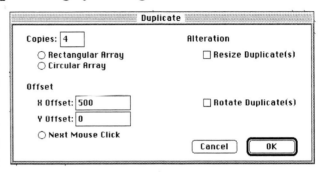

...(Option-Command-D) is used whenever multiple duplicates are required at offsets forming a regular pattern of some sort. This can be:

• Simple (serial): so many copies at such and such a distance horizontally (**X Offset**) and/or vertically (**Y Offset**)

• **Rectangular Array**: choose this and **Copies** is replaced by two fields: **Rows** and **Columns**

(In both these options, negative numbers offset to the left and down, respectively)

• **Circular Array**: choose this and an **Angle** field appears:

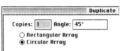

(Remember: positive angles are anticlockwise from 3 o'clock)

In all these scenarios, you have the additional option of resizing and/or rotating the resulting duplicates successively (i.e., each one resized/rotated in relation to the one before it). In Circular Arrays, the assumption is the rotation (if applicable) is by the same angle of the duplication itself; in other cases clicking on **Rotate Duplicate(s) option** asks you for the Angle.

Finally, there is the matter of **Next Mouse Click**. As the name suggests, this tells the dialog to wait, after **OK**'ing it, for your click in the drawing to determine the duplicates' offset (e.g., when you don't know the necessary **X** and **Y**) or—in the case of Circular Arrays—their center of orbit.

Note: *To get back to simple (no-name) duplication after choosing one of the other two, click in the white area above* **Rectangular Array**.

The Reshape Polygon Tool

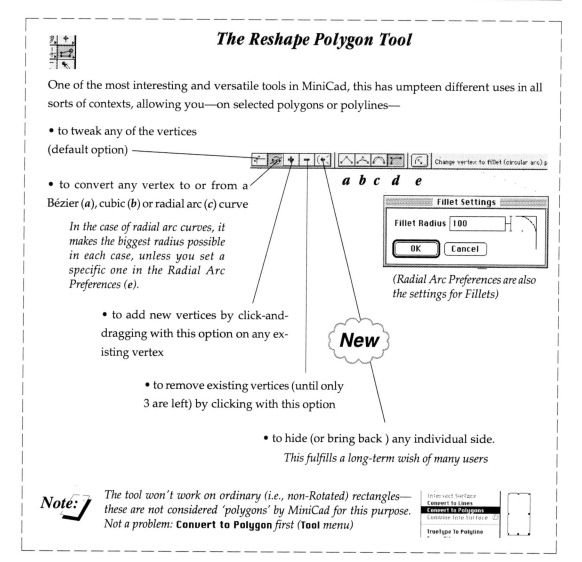

One of the most interesting and versatile tools in MiniCad, this has umpteen different uses in all sorts of contexts, allowing you—on selected polygons or polylines—

- to tweak any of the vertices (default option) —

- to convert any vertex to or from a Bézier (*a*), cubic (*b*) or radial arc (*c*) curve

  In the case of radial arc curves, it makes the biggest radius possible in each case, unless you set a specific one in the Radial Arc Preferences (e).

a b c d e

Fillet Settings

Fillet Radius `100`

[**OK**] [Cancel]

(Radial Arc Preferences are also the settings for Fillets)

- to add new vertices by click-and-dragging with this option on any existing vertex

New

- to remove existing vertices (until only 3 are left) by clicking with this option

- to hide (or bring back) any individual side.
  This fulfills a long-term wish of many users

Note: The tool won't work on ordinary (i.e., non-Rotated) rectangles— these are not considered 'polygons' by MiniCad for this purpose. Not a problem: **Convert to Polygon** first (**Tool** menu)

Intersect Surface
Convert to Lines
Convert to Polygons
Combine into Surface
TrueType To Polyline

Existing Symbols

If you prefer—or just for the sake of variety—you can also use existing symbols that come with the package. The action of placing a symbol from another file has the added bonus of automatically adding that symbol to the internal Resources of the active file, so on future occasions it will be even more accessible.

Call up the Resources palette (Command-R)

If it is merely 'rolled up', roll it down by double-clicking again on its title bar

Click on the pop-up menu near the top of the palette, and drag down to choose the MiniCad package folder

Once there, double-click on the Toolkit folder...

...and then on the AEC folder —and so on inside that until you find the symbol you're after, make it active and place an instance in your drawing

Explore a bit when you have some time—there's lots of interesting stuff, including Commands, macros & hatches as well as symbols

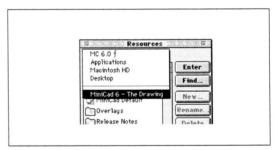

You can at any time navigate down the hierarchical tree to the AEC or whatever level to go up to a different folder

And the program remembers where you're coming from, so you can also at any time return to the Resources of your active file

Once inserted into your drawing, you can of course **Edit** the **Symbol** (*not* **Symbol Edit...**!) to suit your requirements

This is the menu item that normally says **Edit Group** *(but still* Command-[*)*

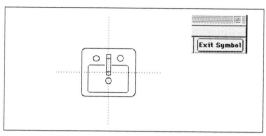

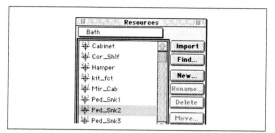

Inside Edit Symbol the intersection of the dotted X and Y axes marks the symbol's Insertion Point. An **Exit Symbol** button is available at the top right. The changes you make affect all instances, past and future, of this symbol.

Bear in mind, however, that unlike the ones we made earlier, some symbols in the Toolkit have both 2D and 3D aspects

Note the little '2' and '3' on either side of the symbols' markers

 These aspects are independent of each other, so by editing the 2D appearance only, you are not affecting the 3D object: something to remember as regards the fidelity of the 3D model to the plan and vice versa. This is something we will return to later.

Meanwhile, speaking of basins—a word from our sponsor:

Tired of Text getting in the way of your details?
Want to make those awkward labels disappear except when needed?
Don't delete them... Don't make a Class for them...

Use MiniCad's great Custom Visibility *facility!*

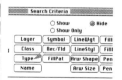

Just click on **Hide**, then on the **Type** button...

Scroll down to **Text**, double-click on it to **Add** it to the list of types of object to hide, then **Done**

OK the Search Criteria dialog and name this Command. **OK** and hey presto!...

A double-click on this Command and all text on the drawing will disappear.

Repeat the process with **Show** *instead to make a* Show All Text *Command to reveal Hidden items*

Stairs

The Straight Stair Tool is a great timesaver and simplifies a procedure that previously involved intricate **Duplicate Array...**s and what not. It automates the task 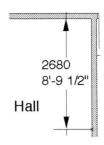 *New* of straight or L-shaped stair runs through a dialog covering pretty much all parameters. You have the choice between 2D only or 3D: we will cover both as part and parcel of our normal procedure

Knowing what parameters to put in requires some planning and measuring, including decisions as to the height to the next level, etc. Assuming we have made these...

 ...Mark the spot where we begin the run by means of the **Locus Tool** (sidekick to Symbol Insert)

> *Use the Floating Datum and the Snap to Surface constraint for this purpose, measuring from the upper righthand corner. One could just as well simply create the stairs first, then drag them into place, but this is as good an opportunity as any to introduce locus points ('loci'), which have no dimensions but can be placed like symbols with one click, snapped to and generally handled like other graphic objects*

2680
8'-9 1/2"

Hall

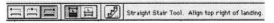 Straight Stair Tool. Align top right of landing.

 Click on the **Stair Tool**, choose the mode of drawing by the right side of the tread, snap to the locus point, and drag up to start drawing.

> *Normally, we would also click on the Stair Prefs. button first, but since we haven't entered any parameters, it appears automatically...*

Click the checkboxes for **2D**, **Square Landing**, and **Draw 2D Stair Break**. Instead of the default values, enter a **Tread** of 280 (11.023"), and **Stair Width** of 1016 (3'-4"). Then **OK**.

Back in the drawing, you now can and should choose the mode of drawing by the front end of the landing (as shown above) before drawing. Click at the locus point and drag up

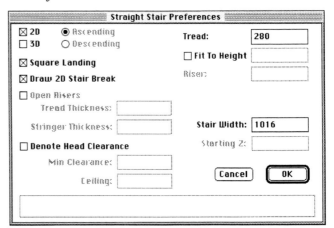

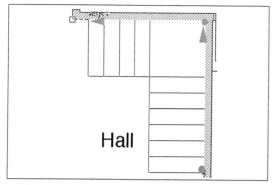

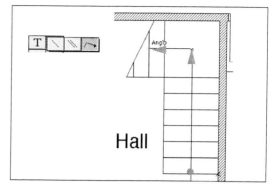

to draw the first flight of stairs.

Because of the Mode chosen, the square landing is created as we go

At the corner, click and turn left, creating another 4 treads before clicking again to stop.

Note the automatic Stair Break made

Aligning up with the center of the first tread, use the **Leader Line Tool** to create the direction indication. Then select all these components (Option-marquee that touches them) and Command-G to make into a Group.

Customizing Arrowheads

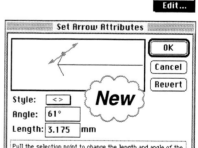

In its default mode, the Leader Line uses a standard solid arrowhead. Click and hold on the two-arrowhead pop-out menu in the Attributes palette to choose another of the standard styles available

Some of them not seen before this version

To change the arrowhead's size and/or orientation as well, choose the pop-out again, this time dragging across then down to **Edit...**

...to call up the **Set Arrow Attributes** dialog familiar to us from the file's **Preferences...** where you have access once again to the standard styles, but can also enter different Angle and Length values, or—even more simply—create the desired arrowhead manually, by dragging the handle in the window.

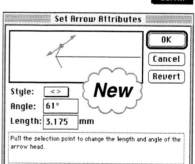

The resulting values are reflected in the value boxes. By the way, any line can have arrowheads added to or removed from it: tick the leftmost checkbox to add one at the start, & the right one to add one

The 3D version of the stair is drawn in essentially the same way, after making the appropriate settings. Its drawback, however, is it does not allow more than one intermediate landing, and ours must have two. To get around this we must create the staircase as two entities: one flight up to and including the first landing, and the two remaining flights with the second landing. As this would create a stair break too high up (only above the passage to the kitchen), we have to be even cleverer. The following describes the issues & solutions arising from use of the tool:

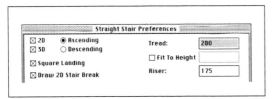

Go into the Stairs Preferences dialog again, and this time click the **3D** checkbox, too, and give the Riser a height of **171 (6.75")**, and **OK**

On the drawing, draw the first flight of stairs just as you did earlier, over the 2D stairs, double-clicking at the landing's upper right corner to complete this section.

Now go back to the Stair Preferences, keep all settings but change **Starting Z** (starting height of the stairs) to **1197 (47.25")**, and **OK**

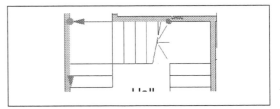

Draw from the upper lefthand corner of the first landing and drag horizontally to the left to the opposite wall, click at the wall, then drag down to create another two (and only two) more treads. Then double-click.

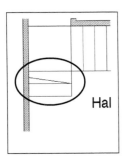

Notice the Stair Break created [*left*]: this is because of the 2D, white-fill, stairs drawn on top of the 3D ones at the same time. They're no good to us here, of course, as it's too high up. Turning off the 2D option removes it, but creates another problem: the 3D object is transparent, which means that you can see the underside joint between the landing and the tread before it [*right*]…

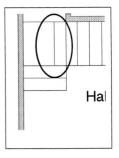

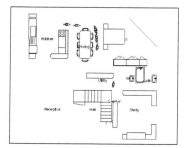

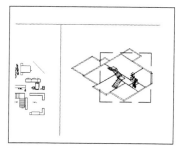

To see what I mean, choose to see the **Right Isometric** view from the **3D>Standard View** menu.

Somewhat confusingly, nothing seems to happen, except that the walls (and other objects with 3D elements) have disappeared.

Zoom out, however, and you will see the isometric of all the 3D elements lying to the right of the drawing.

*This offsetting of 3D views in unpredictable places all around the drawing area—other **Standard** Views result in placements elsewhere—is one of the disconcerting aspects of MiniCad's 3D treatment to newcomers. That, and the fact that items with 3D aspects are apparently removed from the 2D plan. You get used to it after a while. If it's any comfort, the **Projection** views (**3D**)—**Orthogonal**, **Perspective** and the various **Oblique**s— do show the 3D elements anchored to their 2D plans. But they aren't nearly as illustrative.*

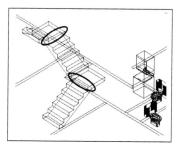

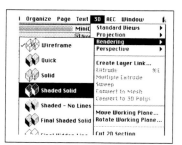

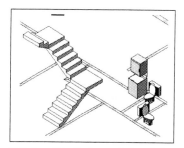

Marquee-zoom in on the area of the stairs and 3D WC items. You can see how the last tread before each landing creates a junction-line underneath it.

To see this more clearly, choose **Shaded Solid** from the **3D>Rendering** submenu. The cursor turns into a little Sun.

Click with it above left of the stairs: the shaded result should appear in a few moments.

*The **Quick** Rendering option is faster and normally almost as good to get a quick appraisal of a model. But it is less reliable in correct sorting of hidden lines.*
*__Save View...__ this view for future use. From here, the **Top/Plan** command (Command-5) can take us back, or better still we can click on our Whole Plan Command.*

Back in the 2D plan, we still have to solve the problem of the stairs' representation. The solution is this:

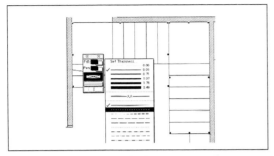

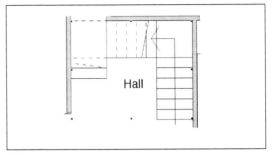

Select both hybrid flights just made, **Group** them (Command-G), **Send** them behind the 2D stairs made earlier (Command-B), and make their lines dashed (from the Attributes palette).

This almost solves it, insofar as the 2D elements of the hybrid flights are now dotted. But not quite, since their 3D components are still of solid line and transparent, which shows where they project beyond the 2D stair break.

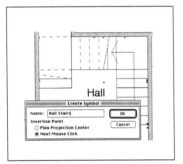

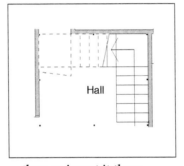

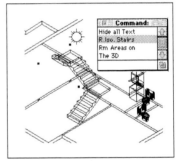

Never fear. The final stage is to select both hybrid and 2D-only stairs and make them into a single symbol—call it **Hall Stairs**. Make the locus point its insertion point...

...then re-insert it there. Miraculously, the 3D parts of the hybrid stairs have disappeared, and only the first 2D stairs and the dotted 2D components of the hybrid ones remain.

And yet (if double-click on the Command made earlier to return to the 3D area) the 3D stairs are indeed there.

If the item is still selected, note the handles, indicating that the 3D projection is one with the 2D representation

How is this possible, you may ask? How did the action of making them into a symbol make the stairs present themselves just as we wanted?

The answer goes to the heart of the meaning and implication of the use of hybrid symbols in MiniCad. This is so important, it is worth emphasizing:

> *When a collection of 2D and 3D/hybrid objects is made into one symbol, they are automatically sorted: the 2D elements to represent the symbol in* Top/Plan, *and the 3D/hybrid ones in all other (3D) views.*

Each aspect of the symbol—the 3D and its 2D representation—can be edited entirely independent of each other. To see this in action—

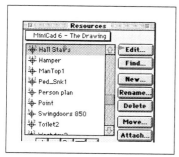

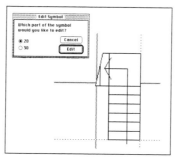

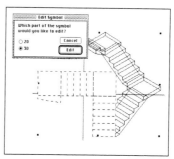

Select (click once on) Hall Stairs from the list in the Resource palette of our file (if your palette isn't up, press Command-R), and click Edit...on the righthand side.

You're given the choice (only if it's a hybrid) between editing its 2D or 3D part. Choose the 2D and you see the symbols drafting representation. Now Exit Symbol (button, upper right), and repeat the process, this time choosing to

Edit the 3D.

In Right Isometric mode, you'll see the above; in Top/Plan, you'll see the two combined in plan as seen prior to the symbol's creation.

The dotted-line 2D drawing is included, because it was created as part of the 3D stairs

The implication of this is crucially important: it means that the *2D drawing is neither limited to, nor lumbered with, the literal appearance of the 3D object seen from above.* If this were not so, we would be severely constrained either in the 2D drafting or in our 3D modelling. The corollary of all this is equally important:

> *In the absence of a dedicated 2D component, the 3D component of the symbol is shown in* Top/Plan, *exactly as it would appear in ordinary (3D)* Top *view.*

(As you can gather, Top/Plan and Top views are not the same: the first is the 2D drafting representation, the other is just that: the orthogonal view of the 3D objects from directly above.)

Understanding these principles helps us greatly in implementing the use of hybrid windows, doors and indeed the walls themselves.

In preparation for adding 3D aspects to the windows and doors, let us first make the walls 3D. Normally, the height would be determined in advance, through a default Z setting for the layer, and in fact we shall do that in forthcoming layers. But, fortunately for us, it can also be done retroactively. To best appreciate the result, return to the Right Isometric view, preferably of the entire plan. Then, select all the walls.

Custom Selection...
"The Key to a Better Life"

Like **Custom Visibility...** mentioned earlier, **Custom Selection...** is designed to make the task of selection particular objects out of a crowd easier, based on criteria that they—and nothing else—have in common. The result, as before, is a custom Command that is added to your Commands palette for future use. The procedure is very similar, too:

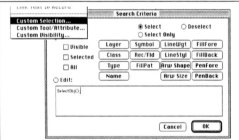

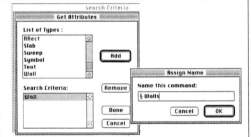

Choose **Custom Selection... (Organize)**, then click on the **Type** button.

Scroll down to **Wall** and double click on it to **Add** it to the list of Search Attributes. **OK** this and the Search Criteria dialog, and name this Command Select all Walls or some such (I use the sign §— Option-6—as a shorthand)

Along with Custom Visibility, Custom Selection Commands offer an easy and powerful alternative to Classes for handling things such as Text and other common categories. A collection of basic Custom-type Commands should be part of any good Stationery file.

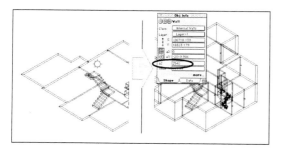
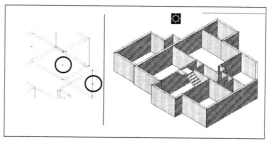

Press `Command-I` to call up their Object Info palette. Since they are all the same object type, there is enough shared information to display. Type 2438 (@: 8") in the ΔZ field, and `Enter`. See the walls spring up.

Note, too, that—in 3D as in 2D—wall sections have only two handles, one at each end.

Now render the scene Shaded Solid as before, to appreciate the space division.

We'll go walkabout a little later.

You no doubt also note that there are no windows or doors in these 3D walls. That is because they have no 3D component. Now that we understand how hybrid symbols work, this is something we can now fix.

In the program's User Manual (p.7-6 onwards), a conventional sash window symbol is shown being created first as a 3D object, then as a 2D representation. Fortunately, we can also come at it from the other direction, i.e., make the 3D aspect after the 2D drawing symbol, which (methinks) is a more typical situation. The following recipe is my improvement on that described in the Manual, but the principle is the same: since the elements of the 3D window are more easily described as extrusions made in the elevation plane, their creation is carried out in Front, rather than Top/Plan, view. The plan of action is to create first the frame, then each of the sashes.

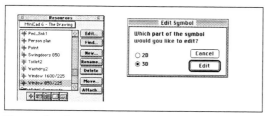
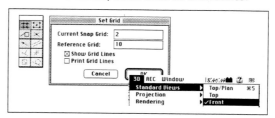

From the Resources palette, choose Window 850/225 (@: Window 3'9"), then click **Edit...** Ask to edit **3D**, as that's the bit of the symbol we're interested in creating

Double-click the Grid constraint to call up the **Set Grid** dialog, turn on **Show Grid**, and **OK**

Without this, we couldn't see the axis lines showing the symbol's insertion point.

Then switch to **Front** view in the 3D menu

Double-click on the Rectangle Tool to call up the Create Rectangle dialog

For situations like these, where we know all the numbers including position coordinates, this 'old-fashioned' method is preferable to the manual-draw-with-Floating-Datum

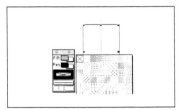
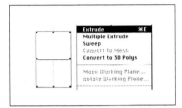

Make **ΔX=850** (2'9), **ΔY=1215** (36), mark the bottom center button as the handle by which it should be placed, and **X=0**, **Y=915** (3'), as its position. **OK**...

...and it should be waiting for you back in the drawing. While it's still selected, make sure it has no fill

This is essential if the final object is to punch a hole through the wall

Click-drag a new rectangle its Top Left to Center Right. Then select the first one, and **Ex-trude** it (**3D** menu) to the wall's thickness (120mm/9").

I choose to do it in this order because when objects become 3D, they lose their center-of-flank snap points. In fact, this can be used as a check that Extrusion took place. Another indication is that it loses any fill it may have—but only temporarily, as it reappears when the 3D model is Rendered.

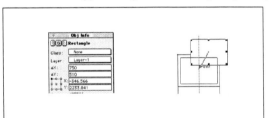
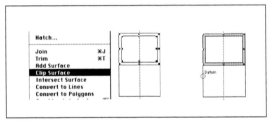

Duplicate the new rectangle, and in its Object Info palette take 4" (100mm) off both its ΔX and ΔY measurements. Drag it by its Center to the Center of its original, then...

Clip Surface (**Tool** menu)—a beep should confirm the operation—then click and delete the inside rectangle to reveal the sash frame.

Check by filling it temporarily with a visible fill.

Draw a single line to represent a horizontal mullion across the inside of the sash frame, then...

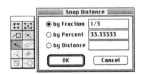

Double-click the **Along Line** constraint, change the setting to **1/3, OK** and draw vertical lines aligning with the 1/3 point of the width of the outside sash frame

(Won't work with the inside of the frame, for some reason)

To cut down on RAM memory requirements and processing time, we will keep the sash frame flat rather than Extrude it. However, to appear in the model at all, it must first be converted into a '3D' kind of flat object. For this, we...

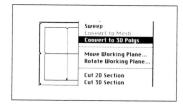

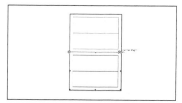

Convert the collection to **3D Polys** (**3D** menu). It blinks, and becomes a single Object.

> *This is confirmed in subsequent Screen Hints*

Mirror Duplicate it with its bottom face as the axis, then **Flip** it **Vertical** to give it the same orientation as the original. Then, select the original one and...

Switch to **Top** view, and see how the sash handles reveal its position. **Nudge** it (**Tool** menu, followed by the cursor keys) a couple of pixels up to offset it from the bottom sash

Because the sash was not extruded, its initial position also revealed the plane from which the frame (the rectangle in the plan) was extruded. This won't do, since the insertion point corresponds with the centerline of the wall, and the 3D window frame would therefore jut half out of the wall. To avoid this, select the frame and **Move** it (Command-M) by a **Y** corresponding to half its depth (in this case, 4.5" or 112.5mm).

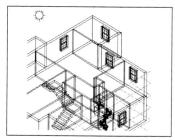

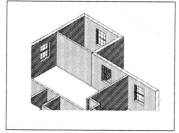

Exit Symbol to see the effect on all the relevant window instances in the model.

Render **Shaded Solid** to appreciate the holes they punch in the walls.

Zoom in to check the appearance of the sashes. Repeat the process for other windows.

 Note: *Our various view changes within the 3D Edit of the symbol referred to the Symbol only: the view for the drawing is as we last left it. Speaking of which, your previous Sun setting is used in all Shading modes, and if you left your 3D scene in Solid Shading mode, rendering is done automatically on your return.*

  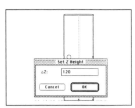 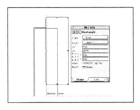

3D Doors

Giving the door symbols 3D components is done in much the same way, and if anything is simpler (see illustration highlights). The main differences are the door leaf is opaque (with the possible exception of top panels in the front door), and they can be shown to be ajar, for greater effect.

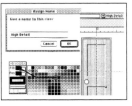

You might like to consider is giving the door leaf a uniform fill of some kind—and perhaps some color, too (from the front rectangle of the Fill pop-out). If so, the effect of inset panels can be made with little RAM or processing overhead by using white-line rectangles, with two black lines added to cover their left and top sides (or vice versa)

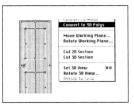

Another good point—made in the User Manual (p. 7-20 onwards)—is the notion of creating High Detail and Low Detail classes for assignment of various components. Except I wouldn't bother with a Low Detail class, and simply use the standard Doors or Doors &Windows class for all except the High Detail items.

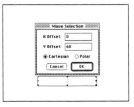

 Note: *Components of a symbol can belong to different classes to that of the symbol itself: this gives valuable flexibility in the rendering of the model later, with the High Detail class turned off except when required, to reduce memory (RAM) requirements and rendering times. All assignment is carried out within the Edit Symbol environment, during or any time after the symbol's creation.*

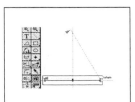

The other issue to bear in mind is that, once a 3D door leaf is added to a door symbol, its 'universal door-swing facility'—i.e. the ability to determine its orientation (in/out, left/right) as you place it in a wall—goes out the window (if you pardon the pun), at least as far as the 3D model is concerned. You will need to make two versions of each door symbol: one Left, and one Right swing.

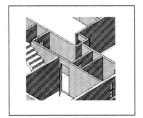

Layers

General

After making 3D components for the other symbols and other basic detail, it's time to move on to other floors. The storey-based structure of buildings leads naturally to the concept of **layers** in all CAD applications, and MiniCad is no exception in this respect. As with the hand-drawn analogy of acetate sheets with which they are often compared, layers offer the benefits of seeing and drawing in registration with other levels of the building, without accidentally moving, deleting, or otherwise affecting objects on them, since drawing operations are typically restricted to objects on your current **active layer**.

What *is* different in MiniCad is that—thanks to its complementary concept of Classes which we saw earlier—layers in MiniCad can be far fewer in number than they are in traditional CAD packages. Thus, there is no need for a separate layers for 1st Floor Structural, 1st Floor Doors & Windows, 1st Floor M&E, etc. Instead, one has one 1st Floor layer, which can be combined with each of the Classes for Structural, Doors & Windows, M&E etc. to create any desired combination for display. This cuts down enormously on the layer count and makes for a considerably 'tighter', more efficient drawing file and easier navigation around it.

Floors (3D)

When utilizing MiniCad's 3D facilities, in addition to having one layer per storey, it is advisable to allocate separate layers to the actual *floorspaces*—the slab containing structural elements, subflooring/screed, and flooring/finish—separating the different levels of a building. This is to avoid the complications that can arise from having to account for the floor's thickness (which may also vary from one part of the building to another) when assigning window sill and heights of other objects in a given layer, which are always measured in relation to the bottom of the layer. We'll appreciate this better in a moment.

To create a new Layer, one accesses the **Layers Setup** dialog. To do this,

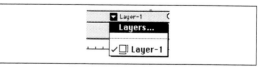

Choose **Layers...** from the **Organize** menu— or better still, from the pop-down in the Mode Bar

Our existing layer is called **Layer-1** by default, and being the only one it is selected in the list to the left. Rename it **Grd Flr Plan**

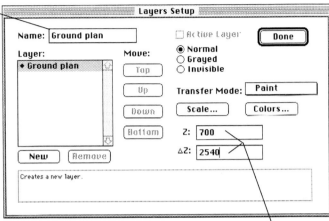

Since the term 'First Floor' means 'Ground' in America and 'the first above ground' in Britain and elsewhere, we shall refer to 'Ground', 'Upper', 'Attic', etc. to avoid confusion.

Secondly, change its **Z** and **ΔZ** settings. **Z** refers to the height at which it begins above a chosen datum; **ΔZ** to the height of the layer itself, which is typically that of its walls. Allowing for our ground floor to be, say, four steps above true ground, enter **700** (27.56") in that field, and **2438** (8') for the Z.

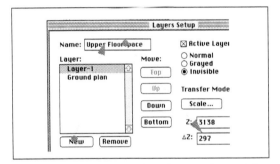

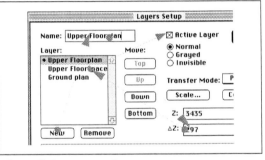

Then click **New**. Note that the new layer (also named **Layer-1** by default, since it no longer applies to the previous one) automatically assumes a **Z** setting equal to the sum of the previous one's **Z** and **ΔZ**. This saves us having to calculate them ourselves, but it also assumes we want the same **ΔZ**, which is not the case on this occasion: change the layer's **ΔZ** to **297** (11.75") and its name to **Upper Floorspace**

Then click **New** again

Alas, unlike previous versions, in this and in other dialogs Command-N *no longer works for this purpose*

Name the new layer **Upper Floorplan**, and give that a **ΔZ** of **2438**(8'). Then choose **Upper Floorspace** from the list and click the **Active Layer** checkbox to 'make it so' (with apologies to Jean-Luc Picard), & **OK**

Note: *If you're not intending to use 3D, ignore the **Z** and **ΔZ** settings and all references to floorspaces, too, and proceed to the discussion of the Upper Floorplan (p.78)*

Back in the drawing, you'll probably now see a blank drawing sheet. This is because the default is to show the **Active** [Layer] **Only**. But we want to be able to draw our Floorspace slab with reference to the Ground floorplan, so choose **Layer Options** > **Show/Snap Others** (**Organize** menu).

*As its name suggests, this not only shows us other layers 'as if we were there', but also allows us to snap to objects therein without the risk of selecting them (for that there is the next and last option: **Show/Snap/Modify Others**)*

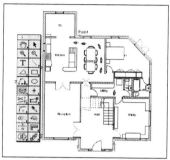

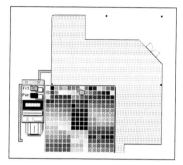

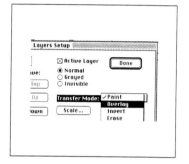

With the **Single Line Polygon Tool**, trace over the ground floorplan without the garage, snapping to each of the relevant vertices in turn. Double-click to close.

Change the fill of the resulting polygon to that of lightly-spaced horizontal lines, and its **Foreground** color (i.e., that of the-pixels-are-normally-black) to a light gray.

Then enter the Layers Setup dialog, select this layer from the list and choose **Overlay** from the **Transfer Mode** pop-down menu. And **OK**.

As you can see, this makes fill patterns in this layer translucent, allowing us to appreciate (and snap to) underlying objects. We can use this to advantage in cutting out a hole for the stairs.

Objects with white fill, incidentally, go completely transparent under this treatment, which is why it's a good idea to use other fill patterns

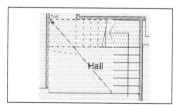

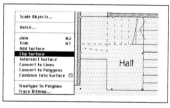

To make the required hole, create a rectangle that snaps between the points shown,

then another one that starts outside it and reaches the point where the stairs are set to end. Use the small one to

Clip Surface the bigger one, then use the resulting polygon to do the same to the polygon of the floor itself

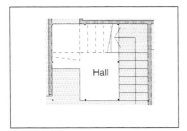

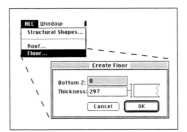

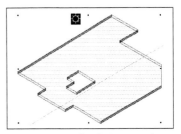

...& remove the clipping polygon to reveal the hole.

Now select the floor polygon and choose **Floor...** (**AEC**). Make it 297mm (10.75") thick, but keep its **Bottom Z** (height of its underside) **0.***

A quick check of the result in **Right Isometric** view should confirm the result.

Note there is no 'fault line' linking the corner of either hole to the perimeter of the floorslab. That's significant, because there would have been had we merely **Extrude**d the floor polygon (as with the door and window frames—try it and see for yourself). That's one reason for using the **Floor...** command. The second is the fact that by this method the floorslab is considered a Group by the program. This makes it much easier to Reshape it, Reshape the hole, remove the hole** or indeed add another one—say, a 'laundry chute' in the floor above the Utilities room:

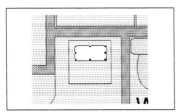

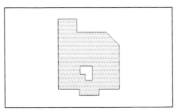

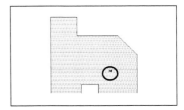

Zoom in on the desired area (in **Top/Plan** view), draw the clipping shape and **Cut** it to the Clipboard (Command-X).

Then select the floorslab and **Edit Group (Organize)** to access the original floor polygon

Note we cannot see the Ground Floorplan—hence we Cut and now...

Paste in Place (Option-Command-V) the clipping shape, select it and the floor polygon, **Clip Surface**, remove the former, and **Exit Group**

---

* You might think it should be 2438 (8')—i.e., on top of the walls of the previous layer. But, as we shall see, for a proper 3D model the objects in each layer must be measured in 'local' terms, i.e. relative to the bottom of their respective layers.

** To remove a hole, simply draw a polygon over it, and **Add Surface** it to the floor polygon

The result can be seen most clearly in Right Isometric or such:

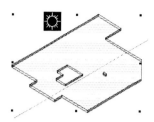

To complete the Floorspace layer, we add three balconies. These will share the same starting **z** height as the internal floorspace, but be of a smaller thickness as they should provide a small drop in relation to the internal floors

> *And because their steel deck-and-concrete construction will give us an excuse to cover such materials in the discussion of details later.*

The different thickness we achieve after extrusion—first we draw the polygons:

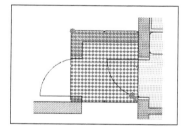

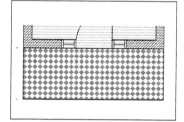

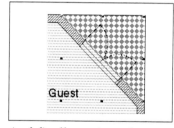

A simple rectangle covering the mudroom/rear entrance to the kitchen

Another one at the front of the house, of the same width as the entrance section and about 1350mm (4'-5") deep.

And finally a triangular one, completing the squarish footprint of the house at the cutoff corner in the back.

Select all three and choose **Floor...** (**AEC**).
Make the **Bottom Z** nil, **Thickness** 186 (7 3/8"), & **OK**.

> *This is of course quicker than applying it to each one individually. But it also means that the three are also made into one Group as a result.*

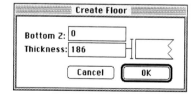

§

Creating an Integrated Model (Layer Linking)

Time to see our two layers together in a single 3D visualization.

Note that each layer can be set independently to a different view: while our Floorspace may be in **Right Isometric** at the moment, the other two are still in **Top/Plan**. One could go into each of the other layers and change them individually, but there is a quicker way:

Align Layer Views (Organize) with that of the one we're currently sitting in. Provided **Layer Options** is set to one of the **Show Others** settings, we will now see a combined view (*right*)

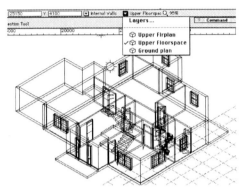

> *Note the now matching icons to the left of the layer names in the pop-down Layer menu*

The depiction is wrong, though: the Upper Floorspace is shown sitting on the same base as the walls of the Ground Floor. This is because they share the same **Z** setting (relative to the bases of their respective layers)—i.e. nil (zero). How do we get the Upper Floorspace be placed at its proper height? Answer: create a new layer, where the objects of all the other layers are 'linked' to produce an integrated model which takes into account each layer's beginning height setting to place it correctly. This is simple enough:

In Layers Setup, create the new layer called (daringly) **Model**, with a **Z** and **ΔZ** of nil so as not to figure in the layer politics. In addition, Shift-select the other layers and mark them **Invisible**: this ensures that these layers will remain invisible even in one of the **Show Others** Layer Options).

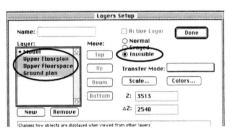

> *This is important as, in order to show the various layers in their correct position, the new layer creates a new image, which will interfere with the images of the individual layers if they are not turned off.*

While you're here, also select the Upper Floorspace layer in the list and change its Transfer Mode back to **Paint** (=opaque)—the better to appreciate the result when the model is Rendered.

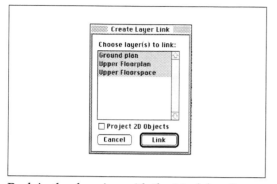

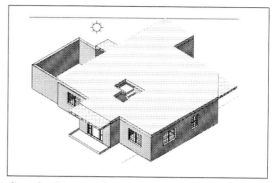

Back in the drawing with the Model as the active layer, ask to **Link Layers...** (**3D**), Shift-select the other three in the dialog, and click **Link** or press Enter.

Even in wireframe, you should be able to see

that the Upper Floorspace is now placed correctly on top of the Ground Floor walls (since its **Z** starts where the Ground walls leave off). A Solid Shade rendering, however, will confirm this unequivocally.

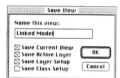

Save this View..., as we will naturally want to return to it frequently. Only this time, ask it to **Save Layer Setup**, too, i.e. replicate the current settings of Model **Normal** (=visible) & other layers **Invisible** each time we use this Command. This will save us from having to do so manually each time.

About (RAM) Memory

This is where memory (RAM) requirements begin to get a little heavy. You may need to increase the memory allocation to the program to complete Rendering the model from this point on.

*To do this, **Quit** (Command-Q) the program entirely and, in the Finder, locate the icon of the actual MiniCad 6 application. Click it once and **Get Info** (**File** menu—or Command-I). Down at the bottom right, replace the Preferred size setting to as high as number as you can spare (measured in K, where 1024K=1Mb)*

To know what that is,

*Consult the Largest Unused Block setting in the **About This Macintosh...** window (menu). Allow at least 1000K for printing (mor e for plotting). If you still run out, consult your dealer about getting more memory. As a partial (software-based) solution, consider using RAM Doubler™, and possibly OptiMem™, too, which effectively double the 'hard' memory installed, albeit with a slight hit in performance.*

*If things getting really heavy, consider using plain **Solid** rendering without Shading.*

New Floorplan

(2D-Only Drafters—rejoin us here)

Now we are ready to create the upper floorplan. Assuming the basic wall configuration has much in common with the Ground level, it makes sense to copy the walls from there rather than create them from scratch (even if the heights are different, as these can be changed easily, as we saw earlier). If you're still in the Upper Floorspace layer, press `Command-[down cursor]` to go down to the Ground level

A useful shortcut. `Command-[up-arrow]` *moves up one layer at a time*

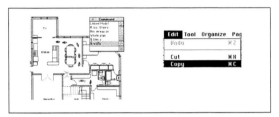

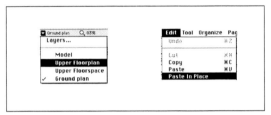

Double click our § Walls (=Select Walls) Command from the palette, and **Copy** them to the Clipboard (**Edit** menu/`Command-C`).

Go to **Upper Floorplan** layer by choosing it from the layers pop-down*. **Paste in Place** (`Option-Command-V`), then...

Go into Layers Setup, mark the **Ground Plan** as **Normal**, **OK**, and set **Layer Options** to **Gray Others**, so we can see the new walls in relation to the existing.

> *An alternative method would be to mark Ground Floorplan as Grayed, then Show Others. The choice is more critical in multi-layer scenarios, where some layers are to be grayed, and others fully visible.*

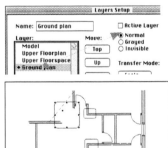

We now do a quick series of editing operations, such as

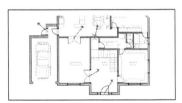

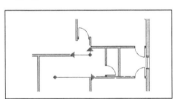

removing irrelevant doors...

(Remember: drag out of the wall before deleting)

dragging back walls in preparation for new junctions...

making new junctions by joining walls...

---

*   I didn't mention it before, but that's the usual way to access a layer. The checkmark of course indicates which one is currently active.

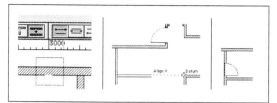

Repairing gaps left by old junctions*, adding new walls where required**, and new doors

Batch-copying symbols from the other level instead of inserting them with the Symbol Tool

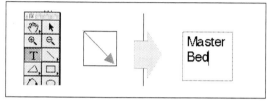

When labelling, have text automatically wrapped by drawing a marquee with the Text Tool to pre-define the textblock's width

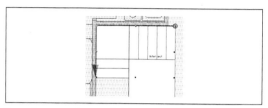

Last but not least, (2D) stairs should be drawn over those of the Ground level (with new, 2D & hybrid ones leading to the next, if necessary)

the result of the walls 'shrinking' proportionally when you resize them

** Since you last created new internal walls your settings for Fill pattern etc. may well have changed and need to be reconstructed. By far the quickest and easiest way to do this—instead of resetting each attribute manually—is to use an undocumented feature of MiniCad's **Eyedropper Tool**.

 Replacing what was in previous versions the menu commands to Pick Up /Put Down Attributes, the Eyedropper's main purpose is to offer quick and easy transferral of fill pattern, line type, thickness, colors, etc. from one object to another. You just click with its tip on the source object (in this case, an existing wall), then choose the Put Down mode (upturned bucket—or press and hold Op-tion) and click on the target. What isn't documented is that even without clicking on a target your settings for any new object are now those of the one you Picked Up from—so just choose your tool and draw. In this respect, this tool duplicates much—but not quite all—of the function of **Custom Tool/Attributes** (of which more later).

Precisely what attributes are Picked Up/Put Down is determined in its Preferences dialog (third button in the Mode Bar).
Note that—contrary to what appears in the program manuals—Picking Up wall thickness and cavities did not make it in the release version. Therefore, if these have changed, you need to reset them manually.

Roofs

1. Sheds

As with Floors and Stairs, MiniCad automates the task of making roofs, particularly of the pitched kind. Like Floors, they are typically placed in a layer of their own, so that the calculations as to their starting **Z** etc. are made easier. Typically, but not always: our first example—that of the roof over the garage—we will in fact create in the Upper Floorspace layer, since its starting height coincides with the top of the ground floor walls. With that layer made active and Layer Options set to **Show/Snap Others**,

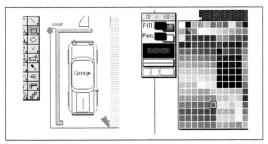
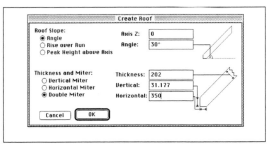

Zoom in on the garage and place a locus point a distance of 350mm (13 3/4") from the upper left corner, aligned with the outside face of its rear wall. Then draw a rectangle from it to the corner of the main structure

> *I suggest using a vertical line fill and, if you like, with the Background color (the default white rectangle) made a desired shade of red or slate.*

With the rectangle still selected, choose **Roof...** from the **AEC** menu, and in the dialog that follows, enter an **Angle** of 26° (**Axis Z**—the height of its support on the wall—stays nil for now). Choose the **Double Miter** option and enter a **Thickness** of 140 (5.5"), a **Vertical** of 90 (3.5") and whatever **Horizontal** results. And **OK**...

Back in the drawing, the **AEC** menu flashes as it waits for you to indicate the line where the roof is supported on the wall—its **Axis line** (*see p. 82*). Click-*drag* (no dainty click-clicks here...) this imaginary line along the outside face of the side wall of the garage [*a*].

An arrow head will appear, pointing right or left as your cursor moves across the page [*b*]. It's asking you to indicate which way lies the ridge of the roof. Make it point to the right, and click.

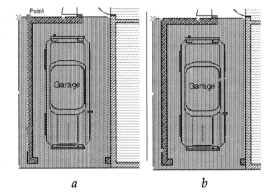

a   *b*

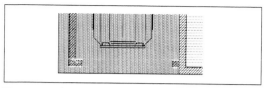

To create the gable end, simply draw a wall at settings matching the one at ground level and snapping to the same width.

Keep it selected and change to a **3D**>**Front** view.

Since all new walls automatically take on the height of their layer, our new addition is, of course, only as high as the thickness of the main floorspace. Fortunately, (as the program manuals point out but is worth reiterating),

The height defined by a layer's ΔZ is not a 'glass ceiling': objects can extend above it. Similarly, its starting height (Z) is not an impenetrable 'floor'

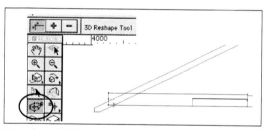

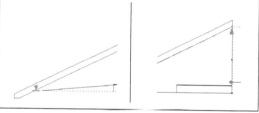

Use the **3D Reshape Tool** (3D tool palette) to grab hold of the wall's left handle and reduce its width to the underside of the roof.

Then drag the top left vertex down to merge with the bottom left one, and the top right vertex up to meet the underside of the ridge.

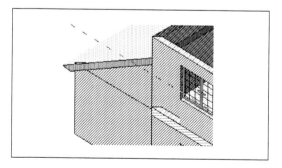

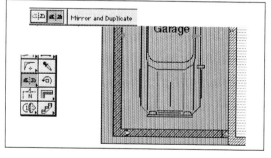

Check your handiwork in **Right Isometric**. If it passes muster, press Command-5 to return to **Top/Plan**…

…and Mirror-Duplicate it to the other end aligning (Shift-dragging) the axis of reflection with the center of the left garage wall.

About (Roof) Axis Lines

The concept of the **Axis Line** of roofs is inextricably bound up with that of the **Roof Plane**. Each is understood in terms of the other.

Basically, a pitched roof's Axis Line is where the rafters sit on the supporting structure. Simple enough, except that this 'sitting point' is not necessarily the underside of the rafters, but rather wherever the roof's 'roof plane' is defined. And this varies in MiniCad according to the type of miter used:

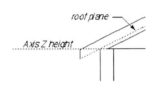

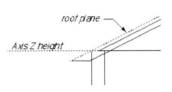

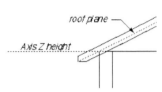

| If vertical, the roof plane is the underside of the rafters. | If horizontal, it's their top surface. | If double (as in our case), it runs through the miter point. |

Since the result determines how high or low the roof sits on the wall, the choice of miter used is particularly important when deciding the Axis Z height in the **Roof...** dialog.

2. Hips, Intersections & Dormers

The main roof works along the same principles, although here, as we said, a special layer *is* recommended, which if you like also incorporates the ceiling of the upper floor (in effect, the Attic Floorspace).

With the Upper Floorplan as our current active layer, click on **New** to create this layer with the correct starting **Z** automatically entered, and with the same **ΔZ** (although it could be made smaller—on this occasion, it doesn't really matter).

A simple gable roof is achieved in much the same way as for the shed/garage to create one half, the result then Mirror-Duplicated to make the other. Where it gets a little more interesting is when it incorporates features such as hips, intersections, and dormers. Let's see how this is done.

The first step—much as in traditional drafting—is to work out in advance the basic roof plan we're after. In order to see the lower layers and to be able to snap to objects in them, we
- **Align Layer Views** to ensure that the other layers are also in Top/Plan mode
- *(in the Layers Setup dialog)* Set the lower floorplans to **Normal** and
- *(back in the drawing)* Set Layer Options to **Show/Snap Others**.

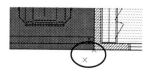

We then place a locus point at the corner of the main structure near the garage, and another at the outside corner of the triangular balcony at the back. Continuing the eaves theme of earlier, place additional loci at offsets of 350mm (13.75")—along both x and y axes—from the first two.

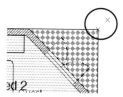

(In case you're wondering, we'll be adding a column to that corner for support)

Draw a rectangle that snaps between them—preferably with a white fill. Then draw single lines from the corners towards the center, constrained to 45° (press Shift), until they overlap. Then use the **Snap to Intersection** constraint (as shown here) and draw another line to represent the short ridge that forms between the hip intersections.

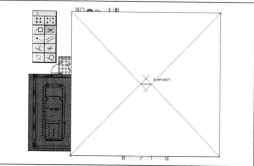

Draw a polygon snapping to the points created to create one of the roof sections, preferably with a suitable patterned fill. The go into the Layers Setup and change the Roof layer's Transfer Mode to **Overlay**, to allow us to see (and snap...) through the fills to the layers below.

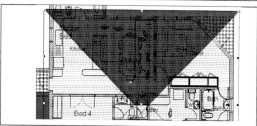

While it is selected, call up the **Roof...** dialog Keep the **Angle** 26° and choose the miter of your choice, but given the bigger spans change the **Thickness** to 190.5 (7.5") and the **Vertical** (if applicable) to 135 (5 1/4")

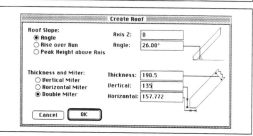

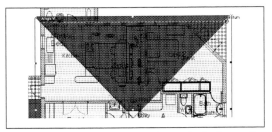

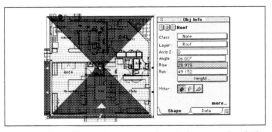

On **OK**'ing the **Roof...** dialog, click-drag the Axis line along the external wall below & set its orientation arrowhead towards the ridge

Mirror-Duplicate it across the ridge, and while this is still selected, call up its Obj Info palette and note its Rise and Run figures

 Note: *This is crucial: unless the other roof sections are identical, simply emulating the same Angle will not result in a perfect joint of the sections at the ridge. Incidentally, I am advised that these Rise and Run figures do not bear any obvious relationship to the actual rise and run in meters or English units, but observing the ratio between them ensures the correct result.*

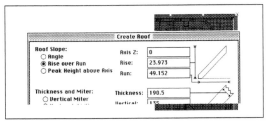

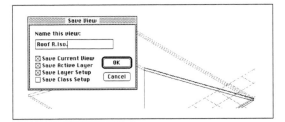

In the **Roof...** dialog for the new section, choose the **Rise over Run** mode and ensure the same figures appear in their respective fields. Back in the drawing, create the Axis line etc.

Keep the new section selected, change to **Right Isometric** view and locate it with double-click-zoom in and out to verify the joints match. **Save View...** for future use

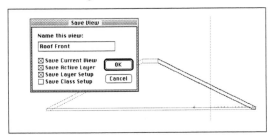

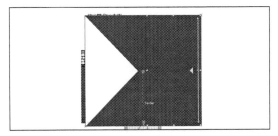

Then check it out in a **Front** view and **Save View...** this, too. Assuming all is well...

...return to **Top/Plan** (Command-5) and Mirror Duplicate it along an axis through the center of the ridge to create the final section.

Intersections & dormers

This is where the fun starts*.

The intersecting roofs of the front portico and at the rear are instructive in the principles of dormers, as well. In each case we apply the **Roof...** routine to a polygon describing one half of the desired addition/dormer. The guiding rule is simple:

> ***Provided the angle of the valley line is 45°, the same parameters that applied to the main roof can be used to create a matching valley line in the new addition.***

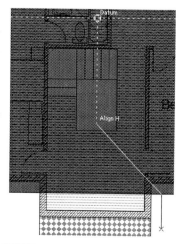

With this in mind, and starting from a locus point at the same offset used earlier from one corner of the front, draw a polygon that meets the main roof line perpendicularly then goes up at 45° until it aligns with the center of the ridge.

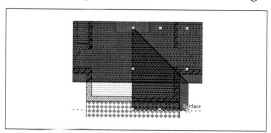

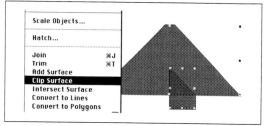

On completion, **Cut** it to the Clipboard (Command-x), select the main roof section and Command-[to Edit it.

Paste in Place the remembered polygon, and use it to **Clip Surface** the main roof polygon. Then Mirror it (without duplication...)

... along the intended ridge and use the 2D Reshape Tool to pull its bottom right handle away from the section already cut before applying **Clip Surface**.

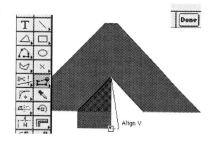

> *Clipping objects that align exactly with one or more of the target's sides tends to confuse the program and can result in the whole of the target object disappearing*

Click **Done** at the top right of the Mode Bar to exit.

---

* What sad lives some of us lead...

Note the main roof section is now cut, ready to receive the new addition.

Paste in Place the polygon again (it should still be in memory). Apply the **Roof...** routine with the same parameters as before—taking care to use the **Rise over Run** mode— and draw the Axis line along the external wall as before, oriented of course towards the center.

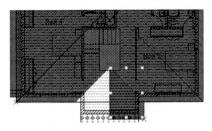

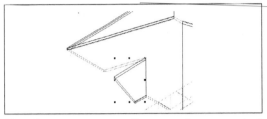

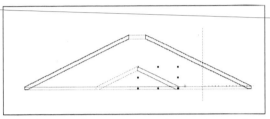

Use the **Save View...** Commands we created earlier to check the result in Right Isometric…

…and **Front**, then return to the **Top/Plan** view. Mirror Duplicate this section to create the other.

Check the result again in Right Isometric—only this time, to truly appreciate your handiwork from all angles, choose the **Flyover Tool** in the 3D palette and, with one or both sections of the new selected, use the first (Rotate about the center of selected object(s)) mode click-and-drag the model this way and that

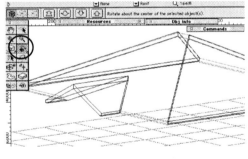

This is one of my favorite tools. The other two modes relate to rotating around the horizontal or vertical planes, and are less intuitive

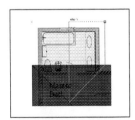
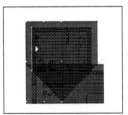
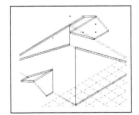
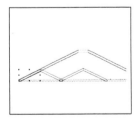

Back in Roof Plan, zoom in on the rear projection and repeat the process there, preferably starting from the section away from the edge, checking the results from the various angles as before.

Assuming all is well, we create the gable ends in the same way as before.

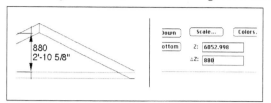

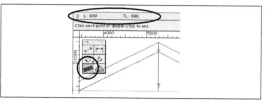

Here, to avoid having to adjust the height of the gable walls after their creation (in **Top/Plan** view, of course), we can measure the rise of the pitch and enter that figure as the new **ΔZ** in the Layers Setup for the Roof prior to drawing the walls (different figures may apply to each gable). Of course, once measured, we don't want the dimension figure staying there, so instead of using the normal dimension tool and deleting the result you might prefer to use the **Ruler Tool**, which simply gives a temporary readout of the distance between the two clicks in the Data Display Bar (with TL showing the cumulative Total Length of the imaginary polyline created by multiple clicks).

 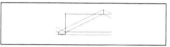 

Having drawn the gable wall at the new default height, it remains only to add-and-drag a new handle to make the peak, reduce the width as required, and drag down the top handles to the base, all as before.

Check the result in Roof Right Iso., or directly in context of the Linked Model in its Saved (Right Iso.) View. Render it.

This may take a couple of minutes, but once it's done this calculation it's done for other angles, too, so getting other views requires only creating a new image. So…

For the first idea of what the building looks to an approaching visitor, change to a **Front** Standard View, and then change the **Projection (3D** menu) to **Perspective**.

For best results, change the walls' fill pattern in all layers to a plain white fill, where the 'white' is changed to a desired color.

On Walkthroughs & Perspective Types

Since we're here, let's go walkabout.

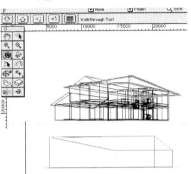

The **Walkthrough Tool** works by pressing the mouse button in a way familiar from many virtual reality applications.

The features are as follows: click-and-hold with the cursor...

- 'north' of (above) the center of the screen to go forward,
- below it to go backwards
- to the left or right to pan in those directions.

Combined with Option key, forward becomes up and backwards becomes down.

Equivalent to the up and down buttons in the Mode Bar

Combined with the Shift key, motion goes into 'warp drive'—enabled by displaying the model as a wireframe block for the duration

The further you are from the center, the bigger the increments and the faster the progress.

Other options are provided by the 3rd, 4th & 5th buttons in the Mode Bar: look down, look up and—very useful whenever you lose your way (which is easy to do)—return to Front Perspective view.

During the pauses, the 'camera lens', too, can be varied: choose between:

Normal *(default)*   **Wide**   or   **Narrow**

Or **Set** your own **Perspective...** . These preset ones are 9.76, 4.43 & 24.41 in the order shown.

What these numbers represent is anyone's guess, but it doesn't matter: in time you learn to gauge by association, like the currency a foreign country by the price of a pint.

Layers for Scale

Having completed the basic fabric of the building, the rest is largely to do with features such as front portico, columns, internal and external balustrades etc. If you are concerned with purely 2D drafting, this involves the straightforward application of circles, lines, rectangles and the like. For a 3D model, however, such items must also have 3D aspects, the heights of which can only be determined through parallel development of a notional section through the building. Either way, we are led naturally to an another very important application of layers, namely for scale.

MiniCad's approach in this matter is different to that of traditional CAD applications. Instead of carrying out everything at 1:1 ('world scale'), with the required presentation scale set only for specific output, each layer can have its own scale, which is that at which it is destined to be printed. This means that one layer can be set to 1:100 (@:1/8"), and another at 1:25 /(@:1/2"), and the two shown together so that a detail and a GA can be shown apparently side by side. At the same time, all objects remain true to their intended size and can be dimensioned at all times for a correct readout.

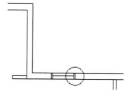

To illustrate this, return to the Ground Plan and draw a no-fill circle around one of the window jambs, such as you might use to highlight a detail. Then **Copy** it to memory (Command-C), and

go into the Layers Setup dialog. Click **New**, zero its **Z** and **ΔZ** settings (it will play no part in the 3D model), give it an appropriate name, and click on the **Scale...** button. This gives us access to the same dialog we encountered at the beginning of the file. Choose **1:25 (@: 1/2")** and **OK**.

The fact that we haven't done so till now tells you that a new layer always takes on the scale of the active one unless otherwise specified

Back in the Layers Setup, make sure our new layer is the new Active one, and that only the Ground Floor layer is Normal, and **OK**.

Choose **Show Others** in the Layer Options to show the Ground Floor plan, and **Paste** (Command-V) the circle in memory in our new layer. Note that it is four times as large as its twin is in the Ground Floor—reflecting the difference between the two scales—giving ample scope to draw the detail, while enabling us still to enter true sizes in creating the components.

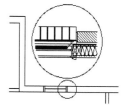

Drawing Details

2D

Up to now, you have been led along by the hand and told what to draw, where, and at what sizes etc., with little or no room for your own imagination. In your own work, however—and indeed for myself as I developed the scenario of this book—I would recommend the use of a separate MiniCad file (**File>New**, or Command-N) to run in parallel with the main project file as a sketchpad to work out details as you go along.

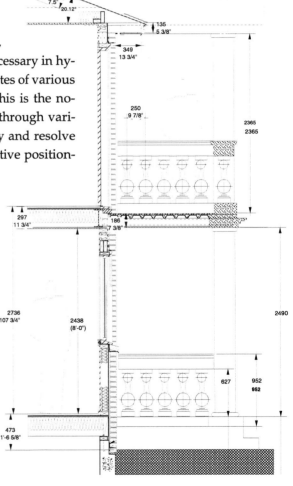

I call this file the 'Jotting Pad' file or some such, and, as mentioned earlier, it is particularly necessary in hybrid files to establish the heights and Z attributes of various features in the model. A good example of this is the notional section seen here—really a composite through various parts of the building—created to identify and resolve issues of wall & floorspace construction, relative positioning of Ground and Upper floors, window and door types, column dimensions, eaves and other details.

Knowing it is a completely separate file allows one to be less precious in fashioning details, creating unfinished symbols or superfluous Classes or layers, and generally to test ideas without fear of impacting on the actual job at hand. In this way one may combine the advantages of rough-and-ready sketching with the confidence that comes with knowing that the results are dimensionally accurate.

It is at this point that one also gets to use some other tools and tricks.

Hatching

You may have noticed that fill patterns stay the same absolute size regardless how far you zoom in or out. This is fine in some cases, but hopeless when it comes to brick elevations and other cases where the spacing needs to be true to the scale being used. This is where true **Hatches** come in, which are different in that they are composed of true lines which are simply drawn in for you by the program to suit the required area.

 A simple but typical example is the hatching of the brick sections. Since many are involved and they will no doubt be required in other projects, it makes sense to create symbols out of each brick type or size, such as the one in our example. The procedure couldn't be easier: draw a rectangle measuring 3 5/8" by 2 1/4" and keep it selected.

Choose **Hatch...** from the **Tool** menu.

> *Note that this option is enabled for use only for selected polygons and other surface-type objects.*

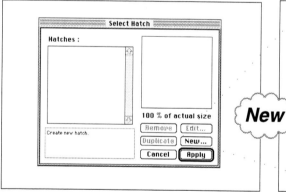

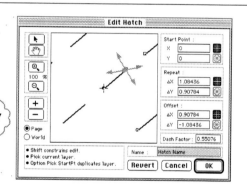

The empty list of hatches to choose from is a little demoralizing…

> *(A good reason for making a basic selection of hatches part and parcel of all your Stationery files)*

and the fact that one can create one's own (click **New...**) doesn't cheer one up all that much, given the daunting plethora of options available (including composing the hatch out of more than one 'layer'—click the '+' button on the left—each one of which has lines going in different directions).

 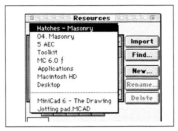 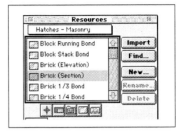

Much better—since hatches are a Resource like symbols—is to borrow (through that palette) from the ready-made hatches available in the Masonry folder in the AEC Toolkit in the MiniCad application folder. Double-click on the Brick (Section) hatch on offer to add it to our file's hatch list, then return to the **Hatch...** dialog to choose it from there.

Even if you find the ready-made variety not quite to your liking, it's psychologically easier to edit an existing one than to create one from scratch, at least until you get the hang of it.

Double-click on our new hatch (same as single click then **Apply**) to select it & return to the drawing, and click with the cursor (now in the form of a tilted bucket, as with Putting Down Attributes) inside the selected polygon. *Et voilà.*

The result can be a little confusing, because although the lines duly appear, one sees no selection handles. This is an illusion: the handles of the polygon and of the Group of lines which is the hatch are cancelling each other out, as they precisely overlap. If you don't like the result and want to delete the lines, make sure you click somewhere first to deselect both, then click on one of the lines to select the hatch by itself.

Note that there is also a reasonable ready-made Concrete hatch available (a tricky one to make).

Structural Shapes

 Another very handy feature is a collection of ready made structural steel profiles. This is accessible from the **AEC** menu, and provides a range of I-beams, channels, angles, etc. in a wide range of nominal sizes.

The resulting profile is an ordinary 2D polygon which can be resized, extruded or otherwise changed, if necessary, like any other.

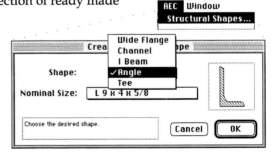

Radial Arcs (by other means)

 Mortar joints demonstrate the use of radial arcs by other modes: specifically the very handy Arc by Two points and Center, which allows you to manipulate by eye the desired center, after clicking on two points representing the arc ends. Although they are tiny, it makes sense to make a symbol for them, since they repeat at every joint (use Option-Duplicate to create as many copies as required at the appropriate offsets).

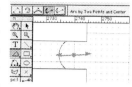

Note: *Insulation, by the way, is another classic candidate for symbolizing, using as the repeating unit the bit of the pattern which fits inside an imaginary rectangle, comprising two lines and three radial arcs (shown here filled, for illustration. Angles may vary).*

*When rotated, however, the sides of the arc segments may suddenly reveal themselves: in that case, **Convert to Lines** the misbehaving arcs and immediately **Group** them for easier handling*

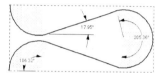

Dimensioning Angles

 Accessed through the Dimensioning tools palette (or by keyboard hotkey '.' [period]), this has two Modes:

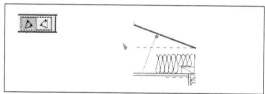

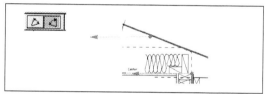

Angle between two objects (default)
where you click on side of the angle (a line or even a polygon side), then on the other, and pull out to place the dimension figure, and…

Angle between reference line and object
where you click-drag an imaginary axis line (often, but not necessarily, orthogonal), then click on the line or polygon side whose angle in relation to that axis you wish to measure, pull out and click to place.

Center Mark Tool

 An stablemate of the **Diameter Dimensioning Tool** in the Dimensioning tools palette, this places a center mark on selected circles, and rounded corners of rounded rectangles, with a single click. Handy.

3D Details

Sweeping

MiniCad's facilities for creating bespoke 3D objects (i.e., other than walls, floors and roofs) are fairly versatile, and having established the relevant heights, Z positions etc. that apply, one should have all the requisite information to hand.

The next step is to consider carefully the level of detail actually needed. Even with the use of symbols, too many or overly-detailed bespoke 3D objects dramatically raises the RAM memory overheads and processing time of the file, making it virtually impractical to view a project such as the one at hand at anything other than wireframe.

A trade-off is called for. One approach is to consider the MiniCad model as a conceptual one, to be exported at this or slightly later stage to a dedicated model/rendering package such as form-Z, Stratavision, Sculpt-3D etc. for greater detail to be added and for high-level rendering for presentations. The other is to create two classes of 3D objects: one High Detail, for use only in restricted contexts showing part of the building at a time, and the other Normal (or just None) for use the rest of the time, to provide just enough to give an impression of the final result.

A good example is the balustrade. Ideally, it would comprise repeated instances of a baluster symbol, created from a lovingly fashioned profile (using the polygon or polyline tool in **Front** view) and…

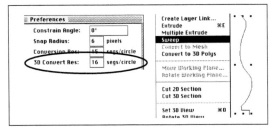 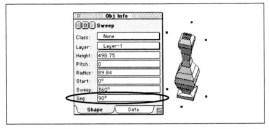

…**Sweep**ed* (**3D** menu—pardon my grammar), at increments previously set in the **Preferences…** dialog (the default is 16)

In practice it is best, at least for Normal Detail, to reduce the number of segments to as few as possible without compromising the design

 Note: As you can tell from the screenshot, this and other aspects can always be changed subsequently through the object's Object Info palette. In addition, the profile of the Swept object can also be accessed and edited by pressing Command- [, just as with Floors and Roofs.

---

* When you want the object to be Swept around a point other than its first vertex, include a locus point in the selection. Oh, and don't include more than one. It's wise to that.

Columns (including capital & base) are also typical objects created through Sweeping (as, indeed, is anything whose profile is more than just half a rectangle). Here, the relatively high number of segments is unavoidable, of course, but try to keep it as low as possible (say, 12).

Resist the temptation simply to **Extrude** *a circle—even if you aren't concerned with capitals or bases. Apart from the greater flexibility provided by Swept objects through the Object Info Palette, Extruded circles constantly change their number of sides to reflect the current* **3D Convert Res.** *setting in the main* **Preferences...** *dialog, as you can see here.*

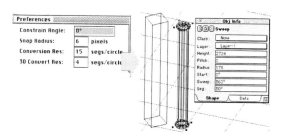

Spiral Objects

Part of the flexibility is that the Sweep Angle can be more than 360°. This, combined with a Pitch that is higher than the profile's Height, produces helixes. The effect is always applied through the Object Info palette after a plain Sweep (try it with a variety of profiles, best with an offset locus). It's a nice feature, but I suspect still more relevant to mechanical or industrial design than to architecture. As a tool for creating spiral stairs, for example, I find it needs too much editing work to be practical.

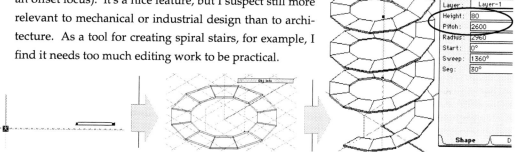

Working Planes

Even with balusters reduced to a modest number of segments, however, the model's processing and rendering overhead may be too high. In this situation, the solution may be to create a balustrade as simple block extrusions in which the semblance of balusters is created by punching through holes corresponding to the spaces between them. Of course, when it's a level balustrade a small wall might be adapted for this purpose, but in the case of staircase railings the extrusion technique is obvious choice.

We have extruded before, so there's no mystery here. The novelty is that on this occasion we're extruding from a profile created in one of the elevation views instead of in plan. This is a good cue for talking about the concept of the **Working Plane**—i.e., the surface on which we are drawing at any given moment. This was less of an issue when we extruded in plan, where the working plane was the Ground or zero plane of the active layer (which is usually what we intended). In elevations, however, the default zero plane is more likely than not *not* to correspond with the surface we would like to extrude from.

Of course, we could always create the thing anyway and move it afterwards into position, but this is an essentially redundant step and is particularly awkward when the target site is at an angle. To this end, MiniCad offers allows up to nine user-defined working planes to be created and saved within any one file.

> *The tenth is the default Ground Plane, which cannot be deleted or renamed. The eleventh plane to be saved replaces the first, and so on.*

To see this in action in a simple but relevant context, let us apply it to the railings of the internal staircase. From the Resource palette, call up the Hall Stairs and click to Edit its 3D component.

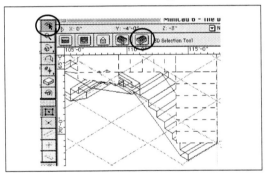

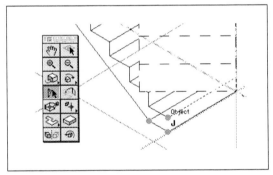

Click on the **3D Selection Tool** (3D palette) and then on the Mode button furthest to the right to view it in Left Isometric

> *This is a shorter method to access the main Standard Views—if you can put up with the dialog*

The thick blue line along the right stringer of the first flight of stairs is the current default working plane in elevation, since the symbol's placement point is there.

To change the active working plane to the left stringer, zoom in on it and, with the **Set Working Plane Tool**, snap-click on any three points on it. The screen will blink, and the blue line will shift to the new position.

> *In Rendered models, a single click on a (filled) surface suffices to indicate the desired plane.*

To save this working plane for future reference,

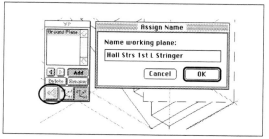

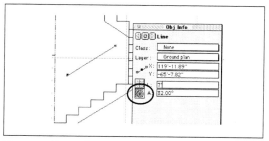

...call up the **WP** palette (**Window** menu), click **Add**, type in a suitable name and **OK**.

Then click on the bottom lefthand button on the palette to switch to a face-on view of this plane.

Draw a line of an appropriate height to indicate the top of the handrail. Use the Object Info palette in its second (**Polar** Coordinates) mode to lengthen it from its middle to cover the whole of the first flight.

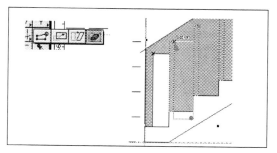

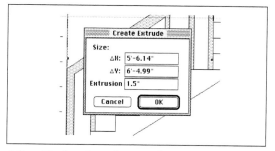

With that line as a guide, create the railing outline as a single polygon, then keep it selected and use the **Clip Tool** to cut out rectangular spaces as appropriate

This works like **Clip Surface***, only with click-drag marquees instead of existing objects.*

After Reshaping the result to your requirements with 2D Reshape Tool, press Option-Command-E to bypass the default Extrusion setting and enter a desired one instead.

This will probably be the way you will usually Extrude from now on, since it gives more control

Multiple Extrude

This is the third principal method for creating a bespoke 3D object. It does so by forming it from a series of 2D elements, each of which serves as the object's section at a particular point. These don't have to be, but in practice usually are, of similar shape or at least follow some kind of evolutionary logic. The order of the sections in the object reflects that of the 'pile': this is usually the order in which they were drawn (last drawn = frontmost = top), unless it was modified afterwards through the use of one or more of the **Send** options (**Forward**/**to Front**/**Backward**/**to Back** – **Tool** menu). In the right context and with some planning, Multiple Extrude can be a powerful tool: the offsets between the sections are admittedly always equal (see below), but as we shall see later this is easily modified subsequently. You could conceivably use it to create certain kinds of column capitals, but a better and less ambitious example of its use in an architectural scenario is the creation of ramps of complex profiles:

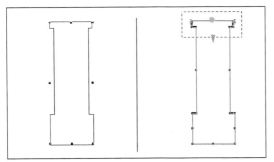
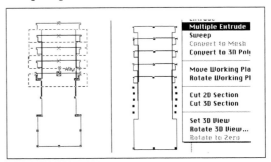

Working in **Front** view, fashion the profile of the ramp at its highest point as a single polygon, duplicate it without offset, then with the Reshape Tool in its default mode draw a marquee around *all* the vertices of the coping and drag that down by its Top Center to maintain its integrity while shortening the midsection*

Repeat the duplicate-and-shorten procedure three or four more times.

A series of locus points duplicated at equal offsets to ensure a uniform slope is a good idea

Finish off by duplicating the shortest profile on itself, then delete the locus points (very important!), select all and **Multiple Extrude (3D)**.

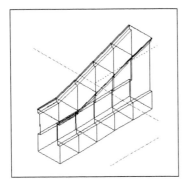
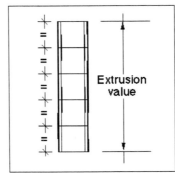
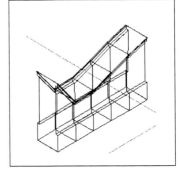

A Right Isometric shows clearly the order in which the sections were created...

The Extrusion value refers to the whole of the object, with the sections being equal parts thereof

...this, like the profiles themselves, can be changed at any time through the normal editing procedure (Command- [ )

---

* This ability to 'batch-move' vertices is an important and oft-overlooked feature of the Reshape Tool.

3D Sections

Whichever way we create our 3D details, there comes the question how
to adapt orthogonally-extruded objects to angles other than 90°, such as
the balustrades of the triangular balcony at the junctions with the house's
walls.

This is where the indispensable **3D Section** facility comes in. Capable of slic-
ing through *any* 3D object in its path (selected or not, Extruded, Swept or otherwise) with the ease of a laser scalpel through custard pudding, it merely asks which bit you
want left behind, and cheerfully deletes the rest. If the prospect of having your lovingly-de-
signed project ending up like the last piece of Mom's apple pie alarms you, be assured that it
carries out its task on a *duplicate*, depositing the result in a new, specially-created layer and
leaving the original object(s) intact.

Nevertheless, to avoid proliferation of such unintended extra layers (a new one is created each
time **3D Section** is applied), I prefer to carry out my 3D Sectioning in a separate file. In the case
of the aforementioned balustrade, assume for the sake of argument a relatively intricate sce-

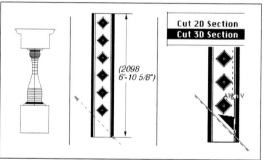

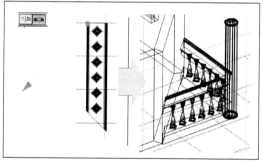

nario where it is initially composed of Swept
baluster instances sandwiched between a base
and coping profiles extruded (in Front view)
to the length required for the long flank. The
collection is then Copied and Pasted to another
file, where **Cut 3D Section** is invoked and applied in a click-drag operation, starting from
the appropriate corner of the Object, pressing
Shift to constrain it to 45°. A final click to
indicate which piece remains…

…and the result is then copied back to the
project file, placed in position, and Mirror Duplicated along a 45° axis as shown, providing
the final result which can then be checked in a
suitable Isometric view.

*You can section even the Linked Model in this
way, too. Remember that the operation is
carried out on all 3D objects along the 'line of
sight' of the click-drag movement, so there is
no need to drag it far.*

2D Sections

Since section drawings are an intended part of the project file anyway, and because they usually involve the whole or significant parts of the building which cannot easily be copied and pasted into other files, this operation *is* suitable for carrying out within the project file.

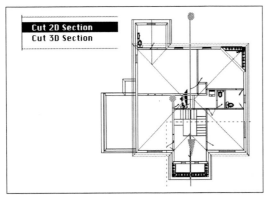

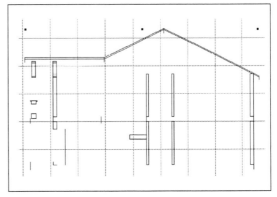

One also works typically in the Model layer in **Top/Plan** or **Top** view, but otherwise **Cut 2D Section** works the same way as its 3D counterpart: you click-drag the section line, and click again to indicate the direction of view.

After the spectacular effects achievable with 3D Sectioning, the result may seem a little disappointing. This is because it is very literally a section and not a Sectional Elevation: nothing is shown of either the structure or 3D details in front of the section plane, 2D objects are ignored anyway

So, too, are Floors, for some reason.

Nevertheless, it usually beats drawing sections from scratch. Don't be misled by the new layer being displayed initially in **Top** view: the objects are entirely 2D, Grouped together. Switch to **Top/Plan** in order to do proper drafting with the various Smart Cursor aids.

Save your file.

§

Drawing Analysis

Graphic Info Scheduling

You may remember—way back at the beginning of the file—we created polygons defining the Room Areas and created a special Class for them. Now is a
good time to bring them back, by double-clicking the **Save View...** Command
that we made at the time to retrieve the combination of Normal and Invisible
Classes that existed before we made the Room Areas class invisible.

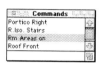

Click on any one—say, the Guest room—and call up its Object
Info palette. All the information you see here is graphic/math-
ematical (**Shape** information, as MiniCad calls it): it is constantly monitored and 'known' to the program, and can be
accessed at any time.

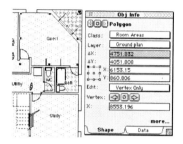

Collating such information into a single Schedule is therefore
very straightforward. We simply define a collection of objects—
in this case, Room Areas—on which we would like a report, in
which the items are listed and the various aspects of its Shape
Information arranged, each in its own column, and the respective values of each item entered accordingly.

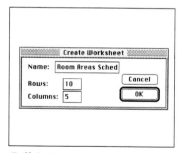

This table-like arrangement is suggestive of a spreadsheet, and
in fact MiniCad's **Worksheet** looks and functions very much
like a spreadsheet. To create one,

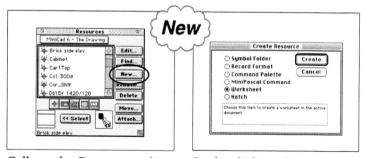

Call up the Resource palette
and click **New** ....

In the dialog, choose **Worksheet** and Enter to **Create**.

Call it **Room Areas Schedule,**
leave the default numbers of
rows and columns and **OK**.

Although it now exists, our new worksheet is still not open. To do so, we must double-click on it from the Resource list. It's down there, buried under the plethora of symbols that we have accumulated: to turn these off temporarily, click on the leftmost button (with the Symbol Insertion icon). The Worksheet is now accessible without scrolling.

As you might guess, clicking on other buttons in that row temporarily turns off
other Resource types, as represented by their icons: worksheets, macros, hatches, and Record Formats

Spreadsheet-like capabilities are nice, but the Worksheet's big strength lies in its hidden and dynamic link with the drawing, enabling reporting and analysis to be done automatically. This kicks in when we define the collection of objects to be analyzed, which we do by designating a particular row to contain the collection (or **Database**, to use its technical term).

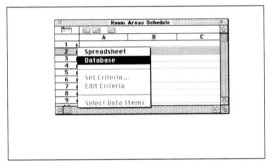

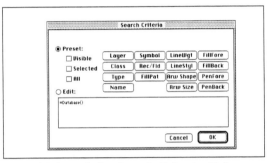

To do this, click-and-hold on the *label* of a row (say, #2, leaving row 1 to accommodate our own text labels) then going across and down to choose the option **Database.**

Our old friend the Search Criteria dialog appears (familiar from the old **Custom Visibility** and **Custom Selection...** days), asking us to define the attributes that distinguish our desired collection of objects. Click on **Class**

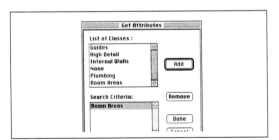

Double-click on **Room Areas** to **Add** it to the list of Criteria, then **Done**, **OK** the Search Crite-

ria dialog. Note sub-rows have sprouted: for each of the items meeting the defined criteria.

The program has already identified the items requested. It will now answer questions about them, posed to it in the form of formulas entered in the 'mother-row' cells.

As in Excel, the entered text will appear at the top of the screen, under the menus

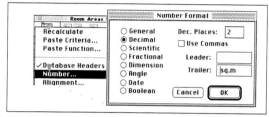

For example, click in cell A2, type =Area and Enter. The areas for each of the items appears, with their total in the 'mother row' cell.

These are <u>in the units defined for the file.</u>
®: If you are using mm as your unit, the result, as shown here, is in sq.mm. To get sq.m., add /1000000 *after* =AREA

To format the numbers to have, say, only two decimal points, choose **Number...** from the pop-down of the small spreadsheet icon at the top left of the worksheet, and choose your desired format and settings before **OK**'ing.

@: For automatic formatting in feet & inches (m'-n"), use the **Dimension** *option.*

Area is but one of the functions that MiniCad understands: =Perim, for example, will give the perimeters of each Room Area. =Length will return lengths of walls and lines. For the (nearly) complete list of functions, or simply to avoid having to type them, use **Paste Function...** from the aforementioned pop-down.

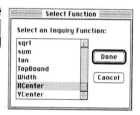

I say 'nearly' as we shall presently be using a very basic one that is not listed.
Remember all Functions need an equals sign (=) typed before them to work—otherwise it's just text. Note that Height *does <u>not</u> mean the z dimension, but merely the vertical size of a 2D object on the drawing.* HCenter *and* YCenter *are useful, though: they provide the coordinates of an object's centerpoint. For an explanation of the others, see section 8 of the program's User Manual.*

This is all fine and dandy, except we still don't know which Room Area is being referred to in each case. The name of an item is not an integral part of its 'Shape' information, so no Function will yield it unless we tell MiniCad what it is in each case (typing labels on it in the drawing in doesn't count—maybe one day).

This is where the flip side of the Object Info palette comes into play—the so-called **Data** side, where we enter information that is entirely of our own making. To see how this works, select the Guest Room Area polygon again and click the **Data** tab in its Object Info palette.

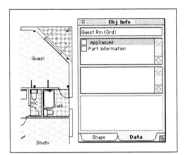

Type Guest Rm (Grd) in the first box in the palette. This is the object's **Name field**, and MiniCad now knows what it's called. Name the other Room Areas by the same method.

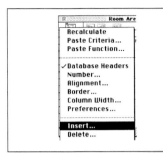

To make room for their names in the worksheet, click on the **A** label to select that entire column and choose **Insert...** from the pop-down to add another one before it

In the new A2 cell, type =N and Enter to produce the corresponding list of Names

The 10 in A2 refers to the number of items that met the criterion

To sort the list alphabetically, click on the **2** label to select the whole mother row, and drag into cell A2 the small icon showing small-to-large

The other sorts from Z to A

As the small 1 in the sort icon suggests, this is our primary sort: secondary sorts can be carried out in other columns

The figures relating the whole collection (not in the sub-rows) can be used in the normal way in other cells for typical spreadsheet-type calculations

But the best bit is when a change is made to one of the items in the drawing: Reshape the Guest Room Area, choose **Re-calculate...** from the pop-down, & watch all the relevant figures change. This is the 'hot-link' in action. It means no more worries about keeping the schedule up-to-date, and to my mind it is one of MiniCad's most significant features as a real-world tool.

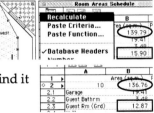

Record Formats & Scheduling User-Defined Info

All of which leads us neatly to the production of schedules of non-graphic data. The name (or D'wg Ref.) that we decide to give an object is the simplest kind of such information. Other aspects might be its price, manufacturer, model & code no., color, material etc. The aspects on which we want reporting—the structure of the information, if you like—can and does vary for different types of object.

Like the label on a shirt in a store, the sum of user-defined information about a given object is known as its **Record**. The structure of the information is known as its **Record Format**. Entering information about a particular object involves first assigning it to the appropriate Record Format, then filling in the values that apply to it specifically.

The Record Formats available for us in the file are listed in the second box of the Data palette. Assigning to them is done simply by selecting the object, then clicking on the Format that applies. MiniCad comes out of the box with two very common types that are likely to be of use to almost everyone: Appliances and Part Information.

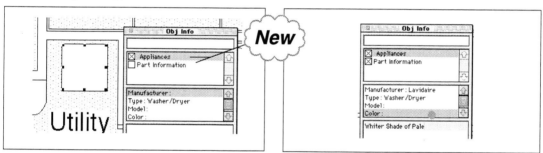

Click on one of one of the symbols you may have imported from the program's Toolkit, for example, and see the list of fields and the values entered.

New or replacement values can be added by simply selecting a field in the list, then typing in the fourth & last section of the palette, and Entering.

Making a new Record Format—say, for Windows—is also easy. Click **New...** in the file's Resource palette to apply for a new Resource of the **Record Format** kind.

> *You can also apply to **Edit...** an existing Format from the list—the procedure thereafter would be the same.*

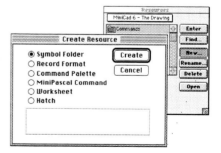

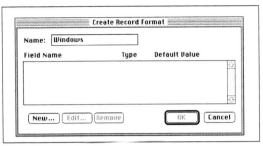

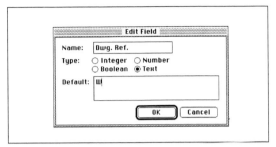

Our proposed Windows Record Format will have the kind of fields we need for a Windows Schedule: Dwg.Ref., Location, Type, Size, etc. These could conceivably be just listed, but to give us maximum control over the formatting and sorting of these fields later on, each of them fields is crafted separately…

…in the **Edit Field** dialog, where we tell it what kind of field we're after each case, and (optionally) the default value, which every object will take on the moment it is assigned to this Record Format.

This will save on typing where all or nearly all items share the same value

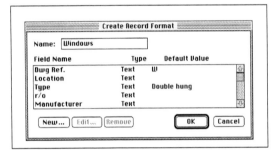

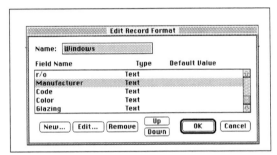

In this manner, create the fields you need for the schedule, and **OK** when finished.

Once created, special **Up** and **Down** buttons are available in the **Edit…** **version** of this dialog to allow you to change the order of the fields

With our Windows Record Format ready, we can now assign the various window instances to it. Instead of having to hunt them or even using **Custom Selection…** in order to isolate them and batch-assigning them (for that, too, is possible), we can guarantee that all instances throughout the file are assigned by **Attach**ing them at master symbol level to the Format. This is done in the Resources palette: select each Window symbol in turn from the list (you cannot Attach more than one master symbol at a time), & click **Attach…**

The advantage of Attaching symbols to a record format at Resource Palette level is most noticeable when many fields have default values, as these are automatically entered for all instances of the symbol once it is Attached. The values for the other fields—or where the default value doesn't apply—are entered manually by selecting the instance(s) in question, selecting the field from the list in the Object Info palette, and typing them in the fourth section, and pressing Enter.

 Note: *The latter is necessary: the program no longer enters it for you after a few moments. However, the fact that the operation does apply to instances already on the drawing is a feature that didn't exist in previous versions.*

 New

Do this now for all the window instances in the file. When you're finished…

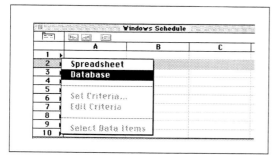

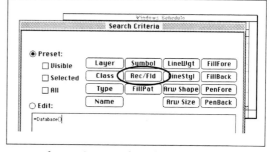

Create a new Worksheet and call it **Windows Schedule**. This time, in defining the criteria for the Database…

…use the main one that the various window symbols have in common, namely their Record Format: click on the button **Rec/Fld**…

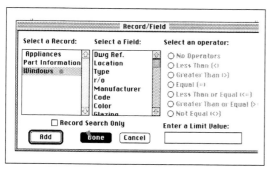

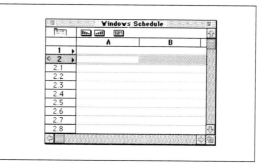

In this dialog, just click on **Windows** from the Record Format list, then **Done**

The window instances have been identified and allocated subrows.

This time, we can't use the =N function, because we haven't entered anything in their Name field. Instead, we will get MiniCad to use the values entered in their Dwg. Ref. field. To do this, select cell A2, type an equals sign, then choose **Paste Criteria...** from the pop-down.

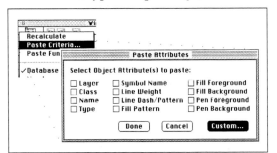

An intermediate dialog appears which isn't relevant to us right now, but note what it does offer for future reference, then click **Custom...**

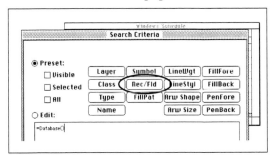

We're back in the familiar territory. Click **Rec/Fld** again...

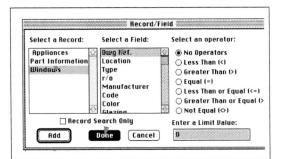

and this time, after selecting the **Windows** Record, click on the **Dwg. Ref.** field from the righthand list, then **Done**

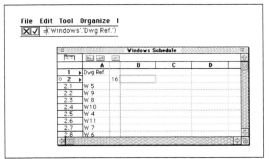

OK Search Criteria, and press Enter to confirm the formula entered for us. The worksheet now lists the Dwg. Ref. values of the various instances, ready for sorting etc.

Using the same procedure varied only by the choice of field at the last step, enter the values for Location, Type, Manufacturer etc. in the other columns of row 2. Use row 1 for typing in reference labels, and congratulate yourself on your first Windows Schedule.

§

More On Worksheets

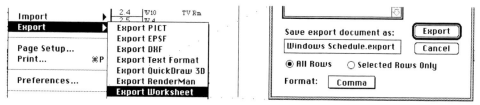

When completed, a worksheet can be **Export**ed (**File** menu)—in whole or in part—in a variety of formats for opening in Excel or other spreadsheets or databases.

If you want to keep it in the drawing, though, various options are available as regards its formatting, both during and after production. These include:

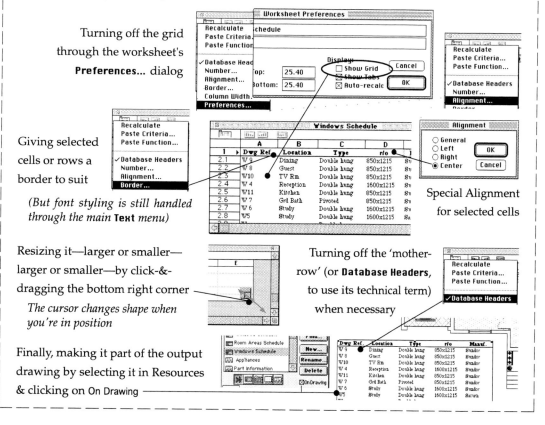

Turning off the grid through the worksheet's **Preferences...** dialog

Giving selected cells or rows a border to suit

(But font styling is still handled through the main Text menu)

Special Alignment for selected cells

Resizing it—larger or smaller— larger or smaller—by click-&- dragging the bottom right corner

The cursor changes shape when you're in position

Turning off the 'mother- row' (or **Database Headers**, to use its technical term) when necessary

Finally, making it part of the output drawing by selecting it in Resources & clicking on On Drawing

Automatic Labelling (Linking Text to Record)

Assigning records to symbol instances has another big benefit. Instead of laboriously typing text labels next to each of the symbols' instances individually, we can have the program do it for us. Staying with windows as our example, use the Symbol Insertion Tool to place a new instance of one (any one) in a blank part of the drawing

Due to the proprietorial attitude of walls to symbols placed in them, we cannot carry out this procedure with such instances, unless the wall has been Ungrouped, which we want to avoid

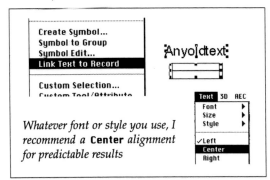

*Whatever font or style you use, I recommend a **Center** alignment for predictable results*

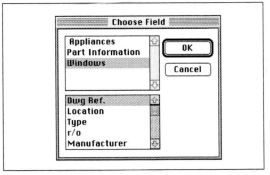

Type a bit of text (doesn't matter what) next to it, in the font, style, and offset you would like applied to all its companions. Then select both, and choose **Link Text to Record (Organize)**

The program asks us which Record (Format) we want to Link to, and which field thereof. Click **Windows** from the top list, **Dwg Ref.** from the bottom one, & **OK**

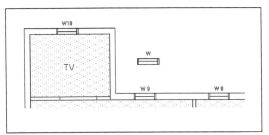

Zoom out and note how all fellow instances of that window symbol have taken on the Dwg. Ref. values assigned to them

Others take on the default value

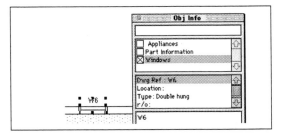

As with the worksheet, the relationship is dynamic: change the value in the instance's Record, and it changes on the drawing as soon as you Enter

As you can see, the facility gives us a 'one-stop-shop' to label all symbols in the drawing, easily and consistently. Since the same information also serves the schedule, it is a real boon.

Out & About

Site Plans

Creating a site plan is typical of situations where one often needs to import an image created elsewhere—in another application or, more likely, on hard copy such as a survey map, existing site plan or the like.

Unless a high degree of accuracy is required—such as provided by an external CAD file or map coordinates—the most common format used in preparing the material for importing is **PICT**, the general-purpose graphic format in the Macintosh environment. In the case of hard copy material, this is typically carried out during the scanning process.

In the absence of a dedicated scanner, have the map faxed to your fax-modem at high resolution.

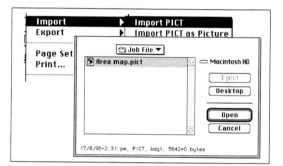

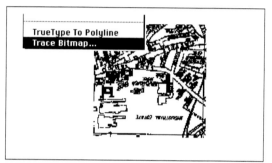

Assuming this is the case, you may **Import** it (**File** menu) through the **Import PICT** option from the submenu pop-out. Locate the file and double-click on it to bring it in.

Select the imported file and choose **Trace Bitmap...** (**Tool** menu). The operation will take a few moments, depending on the original's complexity and the specs of your machine.

Note:

- **Import PICT as Picture** *will not work for our purposes: this makes it unTraceable*

- *With current computer models, this operation no longer presents a particularly onerous memory challenge, or take very long. Nevertheless, to minimize problems, observe the following:*

    — *Cut down the area to be scanned to the minimum necessary*

    — *In the scanning software, 'clean up' the result by removing all superfluous pixel information before saving it as a PICT*

    — *Use a new or 'Jotting Pad' file for importing, and transfer to the drawing file after Tracing and processing*

    — *Set your scale to one similar to that of the hard copy <u>before</u> importing*

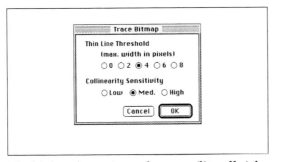

The higher the settings, the more 'literally' the program interprets every smudge and imperfection in the scanned lines. Experiment a bit, but start with the default settings.

Whichever way you play it, the result may well be a bit of a dog's breakfast, unless the original was exceptionally 'clean'. In which case, a manual trace may be a good idea after all.

The exercise wasn't for nothing, however, as—unlike the PICT— the traced file will be visible when viewed from another layer in **Gray Other** [Layer]**s** mode. Which is what we do now.

Also in preparation, choose as our new defaults (i.e., with nothing selected) a thick line and a nice dark blue as our Pen 'Foreground' color

 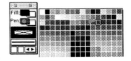

Custom Tool/Attributes

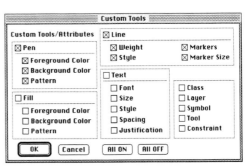

Like its sister Custom… menu items, **Custom Tool/Attributes…** produces timesaving Commands. These are designed to allow you to switch instantly between different 'toolsets', comprising particular settings of pen, fill, line, text…—well, you can read.

It does so by taking a snapshot of the current settings—or at least those which you want recording—for future use. A single double-click on the resulting Command takes you back to these settings from whatever settings may be in place at the time. Another handy inclusion in your Stationery file.

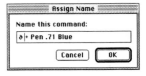

As a shorthand, I use '∂ +' to signify 'default settings plus'

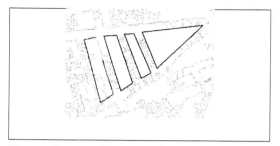

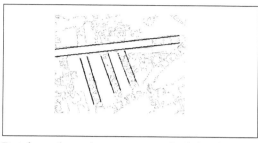

Depending on the type of site it is, you may wish to use a single-line polygon or the Polyline Tool to trace out the blocks…

But for urban sites you may find the double-line tool preferable, as it allows the streets to be defined in short sections…

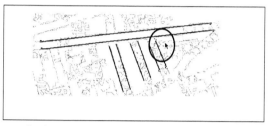

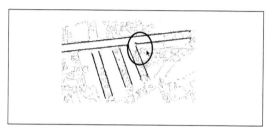

…which can then be joined automatically, by selecting the relevant lines in each junction…

… and choosing **Join** from the **Tool** menu (shortcut: `Command-J`)

The shortcut, incidentally, works with proper walls, too

There remains to reconcile the scale of the imported map with that of the traced image:

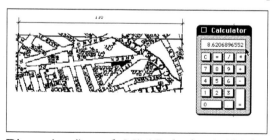

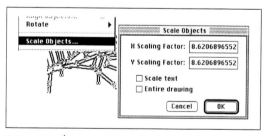

Dimension (i.e., ask MiniCad what it 'thinks' is) a known distance on the original. Then use the Mac's Calculator to work out the ratio by which the traced drawing must be enlarged or reduced by, and Copy the result to memory

Select All (`Command-A`—**Edit** menu) the traced map lines, choose **Scale Objects…** (**Tool**), paste in the result from the Calculator in both the X and Y factor fields, and **OK**. The result should now be ready for transfer to the drawing file.

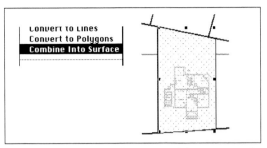

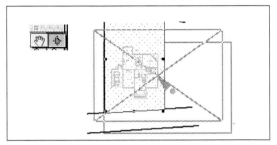

Copied into a layer of its own of the same scale as that of the building, you might **Combine into Surface** the boundary lines of the site itself to allow it to be filled for emphasis

To accommodate the site better on the drawing, you would use the **Move Page Tool** (sidekick to the Pan Tool) to shift the drawing on all its layers in relation to the output sheet

With the Ground Plan set to **Normal** and Layer Options to **Show/Snap Others**, a driveway can be created by **Intersec**ting the **Surface** of a suitable rectangle with that of the site itself, producing the required shape as a separate, new polygon.

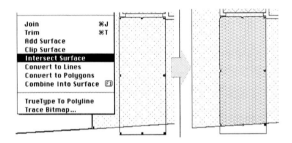

Other design elements may offer opportunities for the 2D Reshape Tool in other modes, such as Change vertex to fillet (circular arc) point (that's 'Round the corners' to you and me)

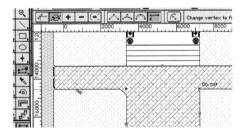

 *Note that the same effect can be achieved with the **Fillet Tool**, which uses the same settings but works differently: instead of clicking on the vertex, you click-drag from one side of the angle to other.*

 The Fillet Tool also offers the option of splitting the lines at the arc endpoints, and/or retaining corner sections. Its biggest advantage over the Reshape Tool, though, is that it works on lines as well as polygon sides—even when these don't actually meet.

 Another likely feature at this juncture is the **Freehand Tool**—great for greenery

But mind how you go: the slower you draw with it, the more handles to the object and the higher the memory overhead.

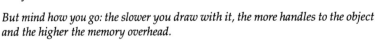

New

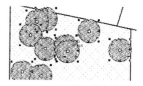

A similar freedom is possible with the **Symbol Paint Tool** (brother of -Insert), which allows instances of the current active symbol to be distributed with broad, painting-like strokes, offering a 'natural', unregimented look appropriate for clusters of trees and shrubs.

Note:

1. If your file is pushing up against the ceiling of the RAM memory allocated to the program, this tool may work very slowly or not at all.

2. It has two modes: one of which is supposed to replace symbols passed over by the cursor (presumably to facilitate 'mixed species' collections), and the other (the default) which doesn't. I haven't been able to discern any difference in operation.

3. Contrary to what is written in the Reference Manual, the Snap to Grid constraint does not have to be on for the tool to work—although on balance it may be advisable.

For a more controlled placement, try the **Duplicate Along Path Tool**—partner to the Mirror Tool. As its name suggests, this allows you to make multiple copies of any selected object—not just symbols—along a path. It has two modes:

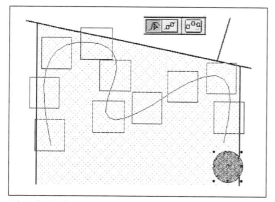

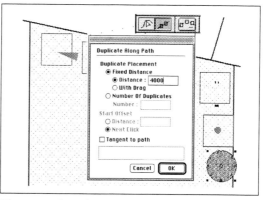

The default—Click on path object and then drag in duplicate direction—varies the distance between the duplicates by the distance you drag after the initial click.

The second option—Drag a polygon path on which to duplicate selection—offers more on-the-fly control, but requires you first to enter the desired offset distance between the instances in the Duplicate Along Path dialog.

Note:

Neither mode requires you to start from the selected object.

As you can see, with the default option the path is not limited to straight-section polygons, but can be curved as well—a big plus when the need arises.

The **Offset Tool** must be *the* dream tool for designers of Japanese pebble gardens, but it has more mundane applications for us ordinary folk, too. It provides parallel offsets to selected polygons, polylines, circles, arcs—in fact, any surface area object. It does so in one of two modes:

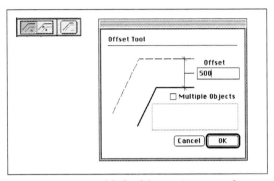

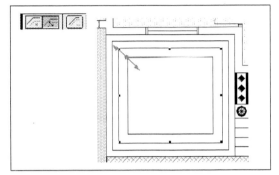

Offset by distance (default) requires you first to enter the desired offset in the tool's dialog...

...while with Offset to a point, you just click and drag to the point you want, with the previous offset acting as the new selected object

 Note: For numerical control the default option is unavoidable, since the program no longer provides a readout of the Offset figure However, it is also more sensitive to memory restrictions.

§

Other Tools & Menus

Part of knowing anything well is knowing which parts of it are important and which less so. The fact that we have come this far and yet there still remains a small collection of menus and tools that we haven't discussed yet is testimony to the depth of MiniCad as an application—but it also demonstrates how much is possible in a typical architectural context without using every single feature available (in other disciplines, of course, the mix of tools and priorities—particularly in 3D—may well be slightly different). I could go on, of course, thinking up ways of accommodating these remaining features in our scenario, but it would feel forced and not truly integral to the plot.

Nevertheless, in other scenarios some of these tools and menus will undoubtedly prove useful if not crucial, so it is worth reviewing them at least briefly, in the order of their appearance in the AEC Overlay:

2D Tools

Quarter Arc Tool
Neighbor to the Radial Arc, this produces quarter-ellipses.

Rotated Rectangle Tool
Devised to save you having to create an orthogonal rectangle and then rotating it. You first draw the base, then drag out the rest.

Regular Polygon Tool

For those situations when only a hexagon/octagon/…*n*-agon will do.
Three modes: Inscribed (center-to-vertex), Circumscribed (Center-to-mid-side), and Edge-drawn (draw a side, and the rest follows), respectively.
The default number of sides is 6—if you want different, change it in its Settings dialog

Trim Tool
Takes a little getting used to, as it works by clicking on *non*-selected objects using the selected object(s) as the scalpel(s). The bit clicked on is removed. Works on lines and most surface-area objects (including circles but not—mysteriously—ellipses).

Extend Tool

From the Trim Tool series, this extends lines to reach a line or object surface. Useful in constructing section and elevation drawings.

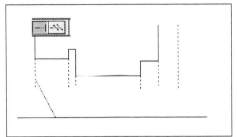 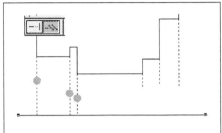

In default mode, you click-drag from the line to the line or surface it is supposed to extend to (neither of them needs to be selected—as in Fillets). In the second mode, the target line or surface is pre-selected, and the lines to be extended to it are clicked on one by one.

Resize Tool

First alternative to the Reshape Tool, this carries out a rapid resizing of any selected 2D object or collection of objects. Different from ordinary resizing in that it's much faster, the object handles are not involved, and anywhere you click first can act as the fixed point.
Someone must have a use for it.

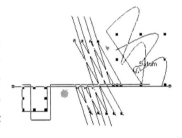

Shear Tool

Third in the Reshape series, this also uses your first click as a fulcrum point, but in this case to skew the selected object(s). Useful in creating the occasional false isometric of details.

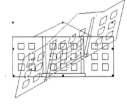

Chamfer Tool

Sister to the Fillet Tool, and works exactly the same way.

Architectural Drafting in MiniCad 6

Property Bounds Tool

When you have precise survey measurements of a site, this tool provides you with the means with which to enter them to recreate the site accurately. Click to mark the starting point, then enter the figures for each line or curve, giving distance, bearing and radius as applicable, and **Apply**ing as you go, until **Done**.

Duplicate Symbol in Wall

This is actually quite an important tool, as it addresses the problem of multiple copies of a symbol placed at equal distances in a wall. Simply selecting, say, a window instance and invoking **Duplicate Array...** wouldn't work, because the wall is selected as well and would also be duplicated.

Duplicate Symbol in Wall solves this by adopting a 'hands-off' approach, similar to that of the Duplicate Along Path we saw earlier. An instance of the desired symbol—*usually not in the wall*—is selected. The tool is then selected, and its settings button clicked in the Mode Bar:

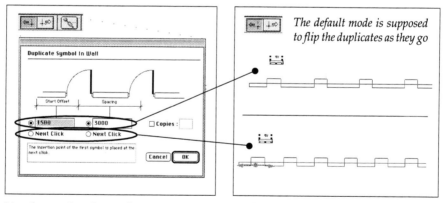

You have the choice between specifying the distances—of the first instance from the wall's end, and of the offsets thereafter—or opting to set both (or either) with your **Next Click** after **OK**'ing the dialog

Back in the drawing, you click on the intended wall section to place the copies; if you opted for Next Click, you have room to manoeuvre before you click again to set.

Number Stamp Tool

A timesaving device to automate the production of serial labelling of construction grid lines and the like. Before use, choose an appropriate default font size and style, and note roughly how big the label shape needs to be.

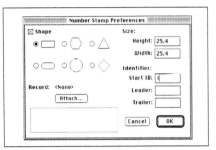

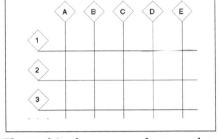

Then click the tool's Mode Preferences button to state the shape and size of the box, and the starting ID. Back in the drawing, click to place each consecutive label.

The tool is clever enough to use letters as well as numbers—but as for the shape's placement point, you can have any you like as long as it's the bottom one. But this is easily fixed.

The reference to Attaching the shapes to a Record in the dialog is intriguing, but its implementation is unclear, since the labels produced are ordinary Groups, not symbol instances.

Revision Tool

Creates a Group of radial arcs to produce a cloudlike shape suitable for highlighting revisions. Between the two modes—by Oval (created like an ellipse), and by Polygon—there is no contest: the latter is slow and produces an unattractive result with an oversized footprint that tends bizarrely also to 'capture' other objects in its orbit.

Fancy Doors

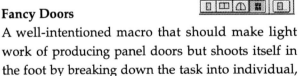

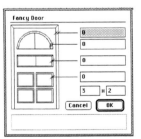

A well-intentioned macro that should make light work of producing panel doors but shoots itself in the foot by breaking down the task into individual, uncoordinated components instead of providing a comprehensive, one-stop dialog for an integrated design. Ah well. Maybe next version.

Shutters Tool

No redundant trips to the Mode Bar with this one: just double-click to launch its settings dialog and set the slat and frame dimensions required. Works simply and well.

®: The default figures are in inches: adjust if you are working in metric

Constraints

Constrain Parallel

With this on, click and drag away from any existing line to produce a line that is parallel to it. If this restriction on where to start the new line is a problem, double-click on the constraint and then click on the target line to make any new line you draw parallel to it until you turn the constraint off.

Constrain Perpendicular

Works just like Constrain Parallel, only to ensure new lines are perpendicular to the target line.

Constrain Angle

Constrains lines and other straight section objects (walls, polygon sides) to 30°, 45°, & 90° increments in any direction. An additional constraint angle can be set in the file's general Preferences… dialog.

Constrain Symmetrical

Works with rectangles, rounded rectangles, ellipses and quarter arcs as they are being drawn or resized manually, to make them squares, rounded squares, circles, and radial arcs, respectively.

In the case of the last two, this means one more way to produce these.

Constrain Tangent

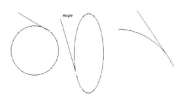

Constrains lines to the tangent of circles, ellipses, and arcs (Radial, not Quarter-) when drawn either towards or away from them.

Press Option *to switch direction on the fly.*

3D Tools

Rotate View Tool

Turns the active part of the screen in a virtual trackball, providing complete freedom in rotating the view of the 3D model in all 3 dimensions, depending on which part of the screen the cursor is. Excellent training for would-be pilots of NASA space shuttles. You can toggle between the modes by pressing u, but as with its stablemate the Flyover Tool, the most predictable results are when you use the first mode, Rotate about center of the selected object(s). The two rightmost buttons turn the view 30° at a time.

Align Plane Tool

I suppose 'Stick This On This' is too informal for a tool name, but it describes its purpose more graphically. It is the answer to the pretty obvious question of how to place one 3D object precisely on the designated surface of another. It works by defining the target surface as the current Working Plane, then clicking on the desired surface (or, in Wireframe mode, on three points defining it) of the object to be placed on it. The second is then placed on the first. Complex 3D objects can thus be assembled out of components which are created individually.

The order in which you click, however, is crucial:

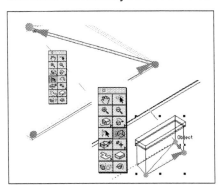

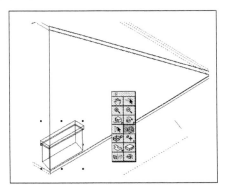

For the most predictable results, define the working surface by clicking on its two base vertices first, and duplicate that order with the object to be placed

This way, the Alignment is made while preserving the object's orientation, which otherwise would not be the case.

Symbol Insert 3D

Works just like its 2D counterpart, though of course only on symbols with 3D components to them. Insertion can take place in any of the orthogonal views (i.e. not perspective).

3D Locus Tool

Since 2D loci don't figure in a 3D model, this serves instead. Is always placed on the current working plane, but its coordinates can be changed afterwards through the Object Info palette.

3D Polygon

As we saw when making hybrid symbols, a 3D polygon is one that may figure in the file's 3D model even though it has no thickness. Useful for creating flat features on 3D objects (e.g. a pediment). Define the surface it is to be created on as the current working plane before drawing.

3D Extruded Polygon

Not, as you might think, merely a 3D Polygon which is extruded at the same time, but a deceptive creator of highly versatile extrusions composed of individually-editable sides:

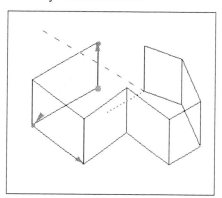

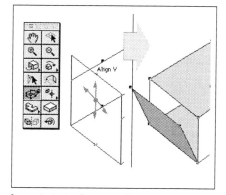

Procedurally, the extrusion dimension is defined with the first click-drag, not the last. Double click to finish. **Edit Group** (Command- [) and see

how it is in fact a collection of polygons, each one of which can, with the Reshape 3D Tool, be resized or even pulled out of the vertical, or deleted, revealing the hollow structure inside.

3D Extruded Rectangle

Produces collections of 3D flat polygons similar to the 3D Extruded Polygon, only rectangular in footprint.

Together, the last two provide the basic tools for 3D massing studies

Mirror 3D Tool

Works much like its 2D counterpart, with similar option to duplicate at the same time or not. The 'mirror', as you would expect, lies perpendicular through the current working plane.

Rotate 3D Tool

Like Mirror 3D, this works only in one of the 3D views (i.e., not in **Top/Plan**), and is analogous to its 2D equivalent. However, it has an additional Mode:

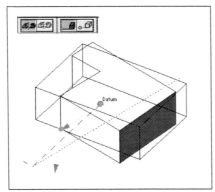

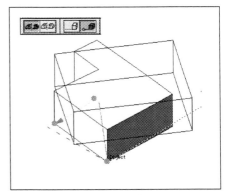

In the default mode—Standard Rotation—it works in the familiar manner: you click to set the center of rotation, drag-and-click out the lever, and rotate (always on the current working plane).

The second mode—Alignment rotation—is a tad more complicated: after setting the rotation lever, you click again (rather than rotate) to set a third point which defines an imaginary line with the center that the object is to align with. The object then jumps to it.

Both modes offer the option of duplicating at the same time, of course

Menus

(File menu)

Revert to Saved: As the name suggests, this takes you back to the last saved version of the file. Used as a last resort when you've tried to fix things but they have gone so badly wrong that you'd rather start again where you left off. Another good reason (if such was needed) to save whenever you've completed something that's gone right.

Import: Apart from **PICT, PICT as Picture**, and **EPSF** which we discussed earlier, the other formats recognized for Importing are:

• **DXF** (Digital eXchange Format): Still the de facto standard for exchange of CAD data (despite its limitations due to the fact that its original purpose is for exchange within AutoCad-based systems), its use with MiniCad is fraught with many issues and considerations which are beyond the scope of this section. For the complete low-down, see Graphsoft's guidelines in the Technical Support section of their Web site (*http://www.graphsoft.com*), or Dave Weber's well-known & updated document *DXF Made Easy* at his home page (*http://www.lookup.com/Homepages/64671/home.html*).

• **Text Format** : Nothing to do with helping you bring specification data into your drawings (for that you use good old-fashioned **Cut-**'n-**Paste**), but a means of reconstructing MiniCad files that may have been corrupted or of versions prior to version 5. You export it from the original file as Text, then import it here. Always use a blank, unformatted file for the purpose.

Also good for importing macros of previous versions

• **Worksheet**: Worksheets in other MiniCad files can of course be accessed through the Resources palette, but this allows you to bring in worksheets saved in common spreadsheet exchange formats such as comma-delimited, tab-delimited, DIS or SYLK. Can be used with active drawing files, but open up a new Worksheet within the file to receive the data into.

Export options cover roughly the same ground as Import, with the addition of:

• **QuickDraw 3D**: Also known as 'metafile', this is a new format developed by Apple for generic exchange of 3D objects between Macintosh, Windows™

and Unix. Includes information on the model's geometry (including convertible measurement information), its lightsource and shading (if applicable). The QuickDraw 3D extension (available from Apple and other sources online) must be installed in your System before you can export in it.

To include a light source in the exported file, render the model with one of the shaded option first. Incidentally, with the QuickDraw 3D extension, you can also store 3D objects in your Scrapbook.

• **Renderman**: Also known as RIB (Renderman Interface Bytestream), this format requires the PixarLibsComp extension to be installed first. Used for exporting to programs such as Showplace™ where it may be rendered to a high level, using cues provided by the patterns and pen colors used in the MiniCad file. Offers special treatment options as regards the shading of individual sides of (true) walls, roofs and floors.

• **QuickTime**: This allows walkthroughs of the model to be saved as a QuickTime movie which may be played back independently for presentation purposes.

The walkthrough path can be either a default 'orbit' path around the object, or a bespoke route that you define. For the former, click the first **Options** button to determine the center of orbit (on Ground Plane, Working Plane, or center of the selected object[s]). A **Preview** button at the bottom gives a very useful idea of the result in wireframe (but only for the Orbit-type, until you've defined a path):

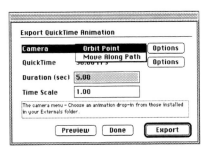

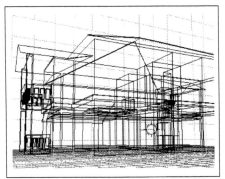 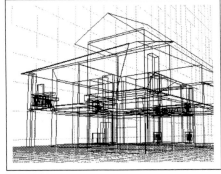

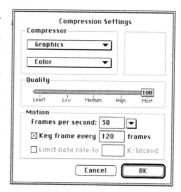

The dialog for bespoke paths offers a comprehensive range of options. The choices you make here are largely governed by the trade-off needed between quality and the size of the resulting file (which can reach dozens of Mbs very quickly): 30 frames per second, for example, is the professional video rate, but rates as low as 15fps are acceptable for most business presentations. The number of frames between **Key frames** (the Saved Views you specify from the list in your file to create a particular **Animation sequence**) also needn't be anything as high as the default 120.

For a full description of the process which is at least as good as any that I could provide, see Chapter 8 in the MiniCad User Manual.

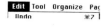

(**Edit** menu)

Paste as Picture does just that, to objects copied to memory (Clipboard). Good for situations where you would like to carry out operations such as **Trace Bitmap...** on an item which isn't in PICT format and so cannot be Imported that way.

Clear—unlike **Cut**—removes the selected item(s) from the file without storing it in memory or Clipboard. Only **Revert to Saved** can bring it back.

Purge Unused Objects... is a housekeeping operation you would typically invoke towards completion of a file or a significant part of it. It removes (per your choice in the dialog that follows) layers or classes that are completely empty and master symbols and/or Record Formats that have not been used.

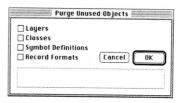

Particularly relevant where the file was launched from a comprehensive Stationery file with the full complement of the symbols, formats, layers etc. of the practice, some of which were not needed. The operation is irreversible, hence you are asked to confirm before it goes ahead.

Edit Attributes... provides a one-stop shop for customizing various defaults settings for:

• **Arrow Heads...**: works like the arrowhead dialog reached through the Fill Attributes palette, only providing for new default sizes and shapes for all types of line markers

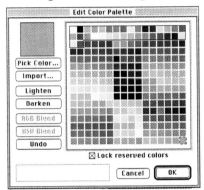

New

• **Color Palette...**: Don't like the range of colors in the color palette?

Make your own, by modifying or completely replacing the color squares with new ones of your choice (even in the reserved selection, if you un**Lock** them)

• **Dash Styles...**: This allows you to modify any of the existing choice of dashed lines in the Fill Attributes, add to it or delete from it. As well as varying the lengths of dashes or the spaces between them, you can add new dash patterns to a given line by click-and-dragging the handle on the right into the mix.

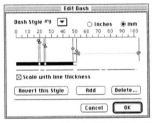

• **Line Thickness...**: A particularly important part of any professional Stationery file setup, this enables you to assign specific pen sizes to each of the five pen thicknesses—in mm, Points, or Mils.

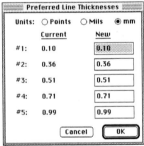

• **Patterns...**: Change the bitmap pattern of any of those from no. 36 onwards in the pattern pop-out—click in the square on the left to change a white (Background color) pixel to a black (Fore color) one, and vice versa. Revert to original if you change your mind. The resulting pattern is seen in the square on the right.

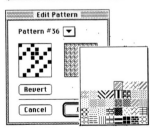

Smoothing | ▶
Lock
Unlock

Smoothing applies to all the vertices of the selected polygon/polyline(s) the effect achieved on individual vertices with the Reshape 2D Tool in its modifying Mode. The result, if applied to a single object as here, demonstrates the difference between Bézier (gray), Cubic (dotted) and Arc curves.

Lock
Unlock

Lock makes the selected object(s) immune to deletion or modification of any kind until it is **Unlock**ed. Useful against inadvertent changes—if you can put up with the persistent reminder dialog that pops up every time you touch the locked item(s).

Tool Organize Page Te **(Tool** menu)

Move...   ⌘M
Move 3D...

Move 3D... does for 3D objects what ordinary **Move...** does for 2D: a **Z** coordinate option is added to that of **X** and **Y**, or you can use their equivalents on the current Working Plane instead.

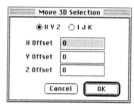

Rotate   ▶
Scale Objects...

Rotate... offers menu alternatives for the Rotate Tool (2D and 3D): dialogs for entering the angle(s), plus one-step commands to **Rotate Left** or **-Right 90°**. In addition, **Flip Horizontal** or **-Vertical** are equivalent to Mirroring on the spot.

Join   ⌘J
Trim   ⌘T

Trim is an operation carried out with a surface object on lines: after selecting both, you choose the command, then click on any of the handles of the surface object. The lines get split along the boundaries of the surface object, ready for removal on either side if required.

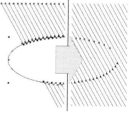

TrueType To Polyline
Trace Bitmap...

Truetype to Polyline is as good as its name in that it converts text of any Truetype font and size (but alas, only in Plain style) to fully editable polygonal objects.

As polygonal objects, the letters (which are Grouped together in the process) can be filled…—

—and even Extruded, if required

Note:

1. For Extrusion to work, letter objects with 'holes' in them will first need these to be proper holes by using them to Clip Surface the bit underneath them

*2. The quality of conversion of curves depends on the **Convert Res** setting in the main Preferences… dialog*

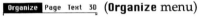

(Organize menu)

Top Level : In 'Russian Doll' situations—where you editing a specific element in a Group within a Group within a Group (or within a Symbol)—this command takes you back in one step to the topmost level, i.e. the main drawing itself. Saves you having to **Exit Group** repeatedly.

Symbol Edit…—not to be confused with **Edit Symbol…**—is an important command that comes into play whenever we need to replace one or more (but not all) instances of one symbol with another.

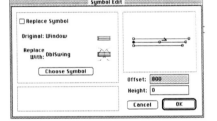

The procedure: Select the instances to be replaced in the drawing, choose the command and, in the dialog that follows, click on **Choose Symbol** to call up the Symbol Library and choose the replacing symbol, clicking on the appropriate insertion point in the right window, and **OK**.

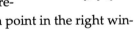

Note:

In the case of wall symbols, instances can only be replaced one by one, due to the impossibility of selecting more than one at a time.

The best results are when the two symbols share the same insertion point and orientation

Set Grid... ⌘8
Set Origin... ⌘9

Set Origin... : Your next click becomes the new (0:0) of the drawing file. Choosing the bottom left of the drawing sheet for this purpose has the advantage that all coordinates in the file thereby become positive.

Guides ▶

Guides: These are unobstrusive, grayedout snap zones, created out of any 2D object. Useful for situations where a special snap pattern is required, without the danger of inadvertent deletions etc. that

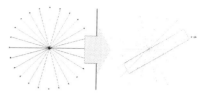

can come about from using ordinary 'live' objects to snap to. Once made, Guides can be deleted, hidden, shown etc. through this submenu.

Tablet

Tablet: Used to map the area of your digitizing tablet (of the ADB kind) to the drawing sheet on screen. You then **Set** the drawing's **Origin** to the bottom left corner of the tablet, and the drawing's (layer's) scale to that of the hard copy to be traced. Then choose this command to start drawing.

(**3D** menu)

Convert to Mesh
Convert to 3D Polys

Convert to Mesh: Actually, this item is too modest. It actually does as good a job as its neighbor in turning selected a 3D object into a Group of 3D polygons that can be individually Reshaped, deleted etc. In addition, however, it redefines the object as a collection of vertices whose X-Y-Z coordinates can be changed even without having to invoke **Edit Group**: manually, using the 3D Selection Tool, or through the Object Info palette. What's more, if you set the

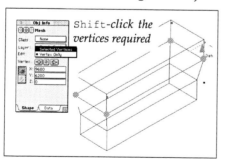

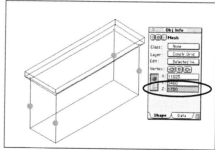

Object Info palette to Selected vertices, you can batch-edit or -drag several vertices at once. The result is best appreciated with **Multiple Extrude**d objects, where the increments were created equal but can now be changed as required.

Set 3D View...: This provides a numerically-controlled method for creating perspectives, using two simple parameters: the height of the observer, and the height of the point being looked at. Typically used when creating the Saved Views to figure as Key frames in exported QuickTime walkthroughs. The positions of the two are the start and end, respectively, of the imaginary line that you draw with the tool when viewing the 3D object or model in plan. This line also launches the dialog.

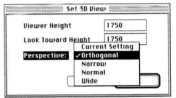

As you can see, you can choose between the three basic types of perspectives. The most natural results, though, are with Orthogonal ones where the viewer and focus heights are the same.

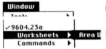 (**Window** menu)

Worksheets and **Commands**—like all other palettes—of the active and other open files are accessed here.

§

Epilogue

or

Things To Do in MiniCad Before Starting

I hope after all the above that you feel confident about tackling your own work using the full range of the tools available in MiniCad. At the very least, you should be aware of how these can help your work should you need them. In the light of what we now know, it may be worth reviewing our first steps in creating the Stationery files that you will use for your actual work. Things such as Drawing Size, Units etc. are trivial, and Scale will vary from layer to layer anyway. What is important is to have the following:

1. A separate, blank, 'Workshop' or 'Jotting Pad' file of the same Units and appropriate Scale settings.

2. Layers: from bottom up:
   — Titleblock layer (typically at GA scale)
   — Construction Grid (optionally combined with Titleblock)
   — Site plan
   — First (ground) storey at GA scale

   > *No point in making others, as their starting heights may vary from one project to the next*

   — Details 1/25 (1/2")
   — Details 1/10 (1")
   — Model (pre-Linked to first storey layer)

No need to produce layers for Sectional drawings, as these will be produced anyway by the 2D Section operation.

3. Classes: for Structure, Existing, Room Areas, Doors, Windows, High Detail, Sanitary, Electrical, F&F, etc. as appropriate for your work

4. Symbols, Hatches: standard practice set

5. Commands:
   — for standard Saved Views (whole drawing, Titleblock area, Detail layer(s)
   — for Custom Visibility of basic types of object: Hide/Show text, loci, etc.
   — for Custom Tool/Attributes: for each of the standard pen sizes and/or different colors

6. Record Formats: for Doors, Windows, F&F and other types

To help in the setup, practise your skills or just for recreation, start with a blank file, set it to a scale of say 1:50 (@:1/4"), set the Wall tool to an appropriate thickness and take it for a walk...

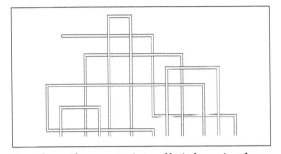 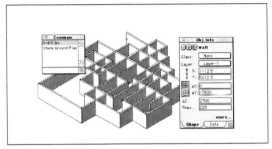

Free from the constraints of brief, zoning laws or conventions, focus just on creating a pleasing composition. Then **Save View...**, change to

a **Right Isometric, Select All**, give them all a ΔZ of some sort, **Render** it **Shaded Solid**, and **Save View** this, too

Go into the Layers Setup dialog, give this layer a name and the **ΔZ** of the walls (or perhaps slightly higher...). Create another layer named **Floorspace**, change its **ΔZ** to something appropriate, then return to **Top/Plan, Align Layer Views** so that the first one does, too, set **Layer Options**

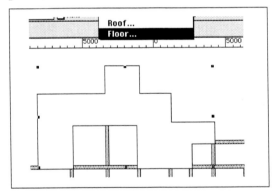 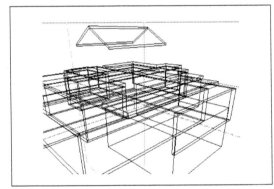

to **Show/Snap Others**, and trace a floor shape over some of the area, invoking **Floor...** at the end of it to give it a depth...

Add another layer, make a few roofs, change the **Projection** to **Perspective**, use the Walkthrough Tool and...—

Shucks...—you know the score, and who am I to teach Grandma how to suck eggs? Just have fun: brainstorming with the fully toolset of MiniCad beats playing *Doom* any day.

If you need more help (training-wise), contact me c/o Qualum. Good luck...

—JS

Index

Y

Z